Introduction to Robotics
In CIM Systems
Third Edition

James A. Rehg
The Pennsylvania State University
Altoona Campus

Prentice Hall
Upper Saddle River, New Jersey   Columbus, Ohio

Library of Congress Catalog-in-Publication Data

Rehg, James A.

Introduction to robotics in CIM systems / James A. Rehg. — 3rd ed.

p. cm.

Includes index.

ISBN 0-13-238395-0 (pbk. : alk. paper)

1. Robotics. 2. Computer integrated manufacturing systems.
I. Title.

TJ211.R43   1997

670.42'72—dc20

95-52303

CIP

Editor: Charles A. Stewart
Production Editor: Mary M. Irvin
Design Coordinator: Jill Bonar
Text Designer: Angela Foote
Cover Designer: Brian Deep
Production Manager: Laura Messerly

This book was set in Palatino by Carlisle Communications, Inc., and was printed and bound by The Book Press. The cover was printed by Phoenix Color Corp.

©1997 by Prentice-Hall, Inc.
Simon & Schuster/A Viacom Company
Upper Saddle River, New Jersey 07458

Printed in the United States of America

10 9 8 7 6 5 4 3

ISBN: 0-13-238395-0

Prentice-Hall International (UK) Limited, *London*
Prentice-Hall of Australia Pty. Limited, *Sydney*
Prentice-Hall of Canada, Inc., *Toronto*
Prentice-Hall of Hispanoamericana, S. A., *Mexico*
Prentice-hall of India Private Limited, *New Delhi*
Prentice-Hall of Japan, Inc., *Tokyo*
Simon & Schuster Asia Pte. Ltd., *Singapore*
Editora Prentice-Hall do Brasil, Ltda., *Rio de Janeiro*

*To the three women
who most influenced my life . . .*

Rose, Marci, and Bettie

Preface

An industrial robot is just another industrial machine. This statement has been used frequently by those in education and industry to calm fears about mass worker displacement or to encourage corporate management to adopt flexible automation. Like the numerical control turning center or the lathe that preceded it, the industrial robot is a machine designed to increase productivity, improve quality, and reduce direct labor cost, but the robot is *not* just another industrial machine.

Robots are versatile: they can be used in every industry that provides goods and services; they can be adapted to numerous job functions; they can change job functions easily; and they work with uncanny skill and unmatched endurance. Robots are different from any industrial machine in the history of automated production. The potential of the robot as a change agent in manufacturing and in our daily lives has not been fully realized.

Although robots are unique, they share one element with other automated production equipment; namely, to be effective they must be integrated into the total solution. Education in robotics must reflect this emphasis on the total system as well. The first edition of this text focused on the robot as a part of an integrated production cell. The interfaces between the robot and cell devices were emphasized. In the second edition, the integration of the robot-automated cell with the computer-integrated enterprise was introduced. Now this edition takes the integration of the robot with the automated cell and system to another level. While robots remain the primary focus of the text, additional emphasis is placed on the hardware and software that support the implementation of automated work cells and manufacturing systems. Major additions to the text include a chapter on work-cell design based on computer-integrated manufacturing (CIM) principles, a chapter on justification of automation projects, and a chapter that describes other work-cell automation used to support a robot installation.

In addition, a major case study with a work-cell design problem is introduced early in the text. The case/design problem continues throughout the text to reinforce concepts covered in chapters and to provide students with an opportunity to design an integrated production cell. Available from Genium Publishing Corp., Schenectady, NY, 800-243-6486, is a videotape of a manual production system, which serves as the starting point for the work-cell design project. Students then follow the progress of an industrial design team as they automate one part of the manual production system. Students are also invited to join in the design process and create an automated cell for an adjacent manual production process. In their solution, students have an opportunity to use skills in computer-aided design and programmable logic controllers. In addition to this major case/problem, the industrial application cases have been expanded significantly.

The book is written for the industrial reader who wants an introduction to flexible automation systems using robotics and for students in two- and four-year colleges. Classroom instruction is supported with questions, problems, and new case projects at the end of each chapter. In addition to the videotape for the case, a disk is provided with the Instructor's Manual that supports the justification calculations and has work-cell models for case projects work. The expanded content of this edition supports a semester course on robotics, and a quarter course can be taught with reduced emphasis on some chapters.

Industrial robots and the concept of a work-cell system are introduced in **Chapter 1.** A brief history of robots is included, along with a rationale for the renewed interest in robot applications in the 1990s. The definition of an industrial robot, a description of a basic robot system, and the new terms used to describe its operation are included along with an introduction to the manufacturing systems that are most appropriate for robot automation. In **Chapter 2,** the different types of robot systems are classified by arm geometry, power sources, control techniques, and path control. The advantages and disadvantages of each type of system are discussed. All of the open- and closed-loop control concepts, which were covered in Chapter 5 in previous editions, are integrated into the classification process in this chapter.

Chapter 3, titled Automated Work Cells and CIM Systems, is a new chapter focusing on the basic process used in the implementation of a single work cell or an entire CIM system. Flexible and fixed automation cells and systems are described and the use of robotics in each is discussed. Two industry case studies are presented as examples of automated system design, and the West-Electric (W-E) automation case is introduced. The W-E case focuses on the automation of successive forging applications in the production of blades for gas turbines and jet engines. In this chapter, the W-E automation project is defined, the W-E automation team is formed, and a survey of the production areas is initiated. End-of-arm tooling is covered in **Chapter 4.** The many different types of grippers are classified into several categories. The wrist interface section, formerly in Chapter 6, is integrated into the tooling section, and coverage of robot tool changers has been expanded. In addition, the section on active and passive compliance includes force/torque sensing in the tooling interface. In the W-E case, a cell is selected for automation and an initial gripper design is begun.

Chapter 5, on automation sensors, describes contact, proximity, and photoelectric sensors. The W-E case focuses on the process of selecting sensors for the automated work cell and the development of robot and work-cell specifications. In **Chapter 6,** Work-Cell Support Systems, the devices and systems traditionally used in automation cells to support the robot are explained. Topics include vision, material handling, part feeding, inspection, and automatic tracking. The vision section encompasses back lighting and front and structured lighting concepts. The W-E automation team works on sensor selection, inspection, and modifications of cell hardware for automation.

Chapter 7, Robot and System Integration, focuses on the control architecture required in the automated enterprise. The major topics are the enterprise network, cell control hardware and software, programmable logic controllers, computer numerical control, servo and nonservo robot controllers, and simple and complex sensor interfaces. The W-E automation team is working on the cell control architecture and on the interface to the programmable logic controller and robot controllers. **Chapter 8** covers the programming issues associated with an automated system. Topics include cell control software, programming the programmable logic controller, robot reference frames, and servo and nonservo robot programming using on- and off-line programming software. The W-E automation team works on programming the robots and developing a control sequence for the automated work cell.

Chapter 9 is a new chapter about work-cell justification and applications. Justification is explained in terms of payback, return on investment, and cash flow methods with an introduction to future value of money and discounting techniques. A spreadsheet demonstrating a discounted cash flow process is included with the instructor's guide. Robot applications are described by means of industrial application case studies. The W-E automation team works on the justification of their automation design.

Chapter 10 discusses the major aspects of the ANSI/RIA R15.06 safety standard for robot automation. The W-E automation team works on the development of a safety plan for the automated cell. **Chapter 11** deals with the problems created by automation with the human interface. In addition, the chapter describes self-directed work teams and how they function.

The robot work-cell design component of the text is integrated into the W-E case study at two levels. The first level is a study of the design process used by the W-E automation team on the automation of the upset forging work cell. The second level is the design of an automated cell to support the extrusion process. Some of the detail is intentionally missing from the W-E team design so that discussion of the case can take place in class. The production information for the entire turbine blade line is provided so that other work-cell projects are possible from the same case. All the W-E case material is provided on a disk in *.dwg, *.xls, *.doc, and *.wpd file formats; therefore, the integration of computer-aid design, programmable logic controller, and other exercises into the case problem is possible.

The instructor's guide includes (1) answers to end-of-chapter problems and case projects; (2) instructions on how to obtain a VHS format tape of the W-E case study; (3) a disk with justification spreadsheets and case study graphics; and (4) work-cell hardware specification sheets.

Many people have contributed to the preparation of this manuscript. I would like to thank the many companies that have contributed material and have been supportive and helpful. A very special thanks to Marci Rehg, Bettie Steffan, and Jerry Bell for suggestions and editing throughout the text. Thanks also to the students and instructors in industrial and engineering technology programs across the country who have provided many suggestions for the third edition.

James Rehg
(jar14@tsu.edu)

About the Author

James A. Rehg, CMfgE, is an assistant professor in engineering at the PennState University Altoona campus. He earned a BS and MS in electrical engineering from St. Louis University and has completed additional graduate work at Wentworth Institute, University of Missouri, South Dakota School of Mines and Technology, and Clemson University. Before moving to Penn State, he was director of the Computer Integrated Manufacturing project and department head of CAD/CAM and Machine Tool Technology at Tri-County Technical College, and previous to that he was director of Academic Computing and the Manufacturing Productivity Center at Trident Technical College. Professor Rehg also served as director of the Robotics Resource Center at Piedmont Technical College and department head of Electronic Engineering Technology at Forest Park Community College. His industrial experience includes work in instrumentation at McDonnell Douglas Corporation and consulting in the area of computer-aided design, robotics, and computer-integrated manufacturing.

Professor Rehg has written five texts on robotics and automation and many articles on subjects related to training in automation and robotics. His most recent text is *Computer Integrated Manufacturing*, published by Prentice Hall in 1994. He has received numerous state awards for excellence in teaching and was named the outstanding instructor in the nation by the Association of Community College Trustees.

Contents

The fact is, that civilization requires slaves. The Greeks were right there. Unless there are slaves to do the ugly, horrible, uninteresting work, culture and contemplation become almost impossible. Human slavery is wrong, insecure and demoralizing. On mechanical slavery, on the slavery of the machine, the future of the world depends.

OSCAR WILDE

Introduction to Industrial Robots

1-1 INTRODUCTION

The introduction of new technology in industrial nations causes changes throughout the social structure of every country in the world. The nature and degree of those changes are proportional to the effects a new technology has on the production of goods and services. Technology in agriculture, for example, caused farming to change from an employer of 80 percent of the population in 1890 to an employer of 3 percent of the population in 1983. Although there was a reduction in the number of farmers, an increase occurred in the number of other jobs required to support the mechanized farm industry.

Today, the application of computer technology in automation is expanding faster than it is in any other area. Networks of industrial computers connect individual manufacturing cells into flexible manufacturing systems (FMSs), and FMSs are linked to form computer-integrated manufacturing (CIM) systems spanning entire enterprises. Although changes in the robotics industry and the introduction of new robot products has slowed from the rapid development pace of the early 1980s, robots continue to have greater change potential than any industrial machine in the past. A study of robot systems and the integration of robots into FMS and CIM systems will be addressed in the following chapters. The first step is to determine what industrial robots are, what they can do, and what they are currently not able to do.

1-2 HISTORY OF THE INDUSTRY

A brief review of robot development is important because it puts the current machines and interest in them into a historical perspective. The following list of dates highlights the growth of automated machines that led to the development of the industrial robots currently available.

1801 Joseph Jacquard invents a textile machine that is operated by punch cards. The machine is called a programmable loom and goes into mass production.

1892 In the United States, Seward Babbitt designs a motorized crane with gripper to remove ingots from a furnace.

1921 The first reference to the word *robot* appears in a play opening in London. The play, written by Czechoslovakian Karel Capek, introduces the word *robot* from the Czech *robota*, which means a serf or one in subservient labor. From this beginning the concept of a robot takes hold.

1938 Americans Willard Pollard and Harold Roselund design a programmable paint-spraying mechanism for the DeVilbiss Company.

1939 Isaac Asimov's science fiction writing introduces robots designed to help humanity and work safely. Three years later he would formulate his *Three Laws of Robotics*, and the view of robots as gothic menaces, presented in earlier works, starts to change.

1946 George Devol patents a general-purpose playback device for controlling machines. The device uses a magnetic process recorder. In the same year the computer emerges for the first time. American scientists J. Presper Eckert and John Mauchly build the first large electronic computer called the ENIAC at the University of Pennsylvania. A second computer, the first general-purpose digital computer, dubbed Whirlwind, solves its first problem at the Massachusetts Institute of Technology (MIT).

1948 Norbert Wiener, a professor at MIT, publishes *Cybernetics*, a book that describes the concept of communications and control in electronic, mechanical, and biological systems.

1951 A teleoperator-equipped articulated arm is designed by Raymond Goertz for the Atomic Energy Commission.

1954 The first programmable robot is designed by George C. Devol, who coins the term *Universal Automation*. The patent application, labeled *Programmed Article Transfer*, was issued as U.S. Patent 2,988,237 in 1961. Devol is joined by Joseph F. Engelberger in 1956; they shorten the name to *Unimation* and form the first successful robot manufacturing company.

1959 Planet Corporation markets the first commercially available robot.

1960 Unimation is purchased by Condec Corporation and development of Unimate Robot Systems begins. American Machine and Foundry, later known as AMF Corporation, markets a robot, called the Versatran, designed by Harry Johnson and Veljko Milenkovic.

1961 The first Unimate robot is installed to unload a die casting machine.

1962 General Motors (GM) installs the first industrial robot on a production line. The robot selected is a Unimate.

1968 Stanford Research Institute builds and tests a mobile robot with vision capability and names it Shakey. Kawasaki Heavy Industries receives a license from Unimation to manufacture the robot in Japan for sale under its brand name.

1970 At Stanford University a robot arm is developed that becomes a standard for research projects. The arm is electrically powered and becomes known as the Stanford Arm.

1971 Japanese Industrial Robot Association (JIRA) is started to promote the use of robots in Japanese industries.

1973 The first commercially available minicomputer-controlled industrial robot is developed by Richard Hohn for Cincinnati Milacron Corporation. The robot is called the T3, The Tomorrow Tool.

1974 Professor Scheinman, the developer of the Stanford Arm, forms Vicarm Inc. to market a version of the arm for industrial applications. The new arm is controlled by a minicomputer.

1975 Robot Institute of America is formed to help US industries effectively implement robots in factory automation.

1977 ASEA Brown Boveri Robotics Inc., a European robot company, offers two sizes of electric-powered industrial robots with a microcomputer controller.

1978 With support from GM, Unimation develops the Programmable Universal Machine for Assembly (PUMA) robot using technology from Vicarm Inc.

1980 The robot industry starts a period of rapid growth, with a new robot or company entering the market every month.

1981 GM brings the Japanese robot manufacturer, Fanuc Robotics, into the US market through a joint venture to develop, manufacture, and market industrial robots in this country. The new company is called GM-Fanuc Robotics.

1983 The robot industry enters a maturing period as industry recognizes that robots and the other automation hardware must be integrated into a unified system. The number of major robot manufacturers falls to twenty-five or fewer.

1984 Direct-drive robot arms are introduced by Adept Corporation with electric-drive motors connected directly to the arms, eliminating the need for intermediate gear or chain drives.

1986	Robot applications and installations continue to grow but with increased emphasis on the integration of the robot into work-cell, FMS, and CIM systems.
1990	ASEA Brown Boveri Robotics Inc. purchases the robotics division of Cincinnati Milacron, and all future robots will be ASEA machines.
1991	The emergence of a global economy, the emphasis on producing competitive products, and use of new microelectronics and display technology in robotic systems spark renewed interest in the robot in integrated automation systems.
1992	After helping to make Fanuc Robotics the sales leader in the United States (annual sales of $200 million in a $700 million market), GM sells its interest because of financial troubles.
1994	Robot sales in the United States for the first 6 months set new records in number of units sold (4,355) and in total value ($383.5 million). The two largest companies were Fanuc Robotics Corporation and ASEA Brown Boveri (ABB) with annual sales of about $200 million and $120 million, respectively.

In the beginning, the domestic robot industry was dominated by US manufacturers (Unimation, GM, Cincinnati Milacron, Westinghouse, IBM, General Electric, Prab, Brinks, and others). Today, the major manufacturers of robots used by US industries are international and include Fanuc Robotics Corporation, ABB, Motoman, Panasonic Factory Automation, Kawasaki Robotics, Sony Component Products, and Nachi Robotic Systems. All of these companies are Japanese except for ABB, which is a Swedish industrial conglomerate. Adept Technology, Inc. is the largest US-based robot manufacturer with annual sales of $50 million. Japanese dominance results from the strong adoption of robot technology by their domestic industries. The US Robotic Industries Association estimates that there are about 300,000 robots in use in Japan compared with 52,000 in the United States. Automotive companies are the largest users of robots, but the demand for robot automation in other industries is growing. The primary driving force behind the renewed interest in robots is lower unit cost and greater reliability; for example, today's prices for robots are half what they were 10 years ago for comparable machines, and reliability is four times better.

1-3 THIRTY-YEAR-OLD INDUSTRY

As our review of history showed, robots are not new; in fact, mechanical components of robot-like devices date back to the early part of this century. Robot technology, available in 1960, was not adopted in large scale in the United States until the late 1970s. There are two reasons for the delay in the application of robots in US industry.

The first reason relates to *economics* and the second to *hardware*. In the late 1960s and early 1970s economics was the primary barrier preventing the rapid deployment of robots into industrial applications. The return-on-investment (ROI) policy in most US industries required that new machines pay for themselves in 1 to 2 years. Early robot applications focused on replacement of human labor; as a result, direct and indirect labor costs were the primary component used to calculate the ROI. In the 1960s and early 1970s the hourly cost of the robot equaled or exceeded that of a human operator; therefore, the robot could not be justified economically. In 1960 the cost of operating a robot was more than $9 per hour, whereas the overall cost of a human operator was less than $5 per hour. Figure 1–1 shows that the hourly cost advantage switched to the robot in the early 1970s when the United States experienced several years of double-digit inflation. In addition, the cost of human labor increased at that time because of enhanced benefit packages and health plans. This dramatic increase in the cost of human labor in the mid 1970s started the trend to use robots in manufacturing and established the industrial robot industry. Roger Smith, the chairman of GM, remarked, "Every time the cost of labor goes up a dollar an hour, a thousand more robots become economical." The demand for robots in the 1980s would have continued to rise if it were not for a significant change in the labor picture midway through the decade. The yearly increases in labor cost were

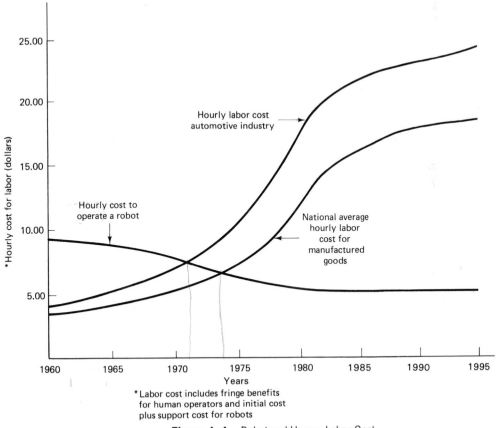

Figure 1–1   Robot and Human Labor Cost.

small across most of the manufacturing sectors and dropped in some areas. The reasons for leveling in the hourly rate included a reduction in the demand for labor as a result of increased use of automation, implementation of more efficient production systems, downsizing of the manufacturing and service workforces, and little growth in the price of basic resources and raw materials due to noninflationary fiscal policies in the major manufacturing countries.

The second factor that influenced the emergence of the robot industry in the late 1970s was the development of the *microprocessor,* or computer on a chip. The robots available in 1970 had remarkable capability, but their versatility pales before that of machines currently being offered by robot manufacturers. The primary difference is not mechanical but electronic sophistication. The microprocessor, developed by Intel Corporation in 1971, provided robot manufacturers with a source of very inexpensive intelligence to drive their machines. The first computer-controlled industrial robot, the Cincinnati Milacron T3, used a single minicomputer to direct all the robot motion and machine control within the work cell. The machine had tremendous capability but was limited by the speed of its single computer, which had to do all the calculations. New robot controllers use a network of powerful microprocessors to simultaneously solve the many concurrent problems associated with robot motion and control. The addition of microprocessors and improved software provided intelligence and operating speed not found on early robots.

Industrial robots became a significant production tool in the late 1970s and early 1980s due to increased labor cost and the development of the microprocessor. However, the introduction of robotics in industry slowed in the mid-1980s as a result of the leveling of labor cost and a greater emphasis on the development of integrated production systems. Robots proved that they could satisfy unique manufacturing problems, but robots could not be stand-alone solutions because industry was moving toward integrated manufacturing cells and production systems. This move to integrated solutions was driven by pressure on companies to meet a host of external and internal challenges.

1-4 INTEGRATED SYSTEMS—MEETING THE EXTERNAL AND INTERNAL CHALLENGES

To remain competitive in global markets, US manufacturers must recognize and meet two challenges, one external and the other internal. It is important to understand the nature of these challenges facing manufacturing, because the challenges drive the use of integrated manufacturing systems and the selection of specific robotic solutions.

External Challenges

External challenges result from a variety of forces and conditions outside the enterprise. The challenges illustrated in Figure 1–2—*niche market entrants, traditional competition, suppliers, global economy, cost of money,* and *customers*—are common for most global manufacturing enterprises. Let's examine each of the items in more detail.

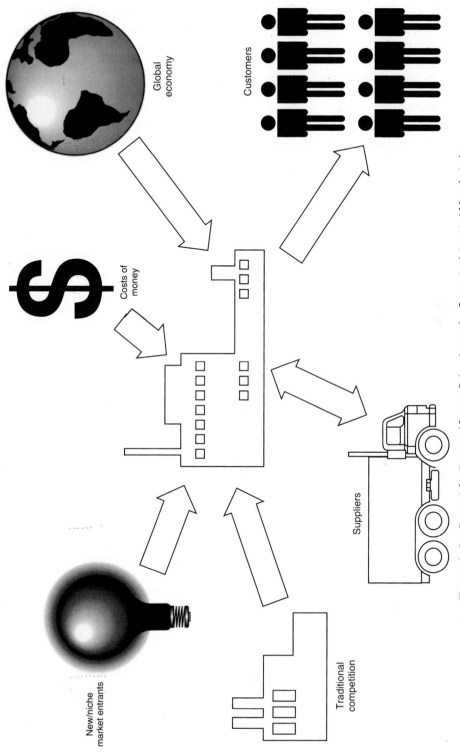

Figure 1–2 External Challenges. (*Source:* Rehg, James A., *Computer Integrated Manufacturing.* Englewood Cliffs, NJ: Prentice Hall; 1994, p. 5)

Global economy

Customers

Costs of money

Suppliers

New/niche market entrants

Traditional competition

Companies have always had product lines called their cash cows. These are high-revenue-generating products that carry the rest of the product line. The success of the cash cow is a result of market dominance, technological superiority, patent protection, or the absence of competition. IBM's mainframe computer line in the late 1960s and early 1970s is an example of this phenomenon. The production and design technology available today, however, makes it possible for small companies to develop a product in a *market niche* and compete with an established producer. In the IBM example, a former employee recognized the profit potential of the mainframe computer market niche and set up a company, Amdhol, to compete with IBM in this area. The Amdhol computers could run IBM software and work with IBM peripheral devices, so only high-profit computers were sold without the need to develop application software or supporting equipment.

Traditional competition, the second challenge, continues to be a problem for manufacturing. Manufacturers often counted on the lack of technology and efficiency of competitors as a buffer for their own shortcomings. Those days are gone. The technology to increase market share is affordable and available to everyone, so the challenge to stay ahead of the traditional competition is stronger today than ever.

Another external challenge is present in the *purchaser-supplier relationship.* In the 1980s many companies purchased parts and subassemblies from vendors with the lowest bid. Price continues to be a factor; for example, a study by a midwest injection molding die manufacturer indicates that a purchase price that is 9 percent higher than the competition can result in a 22 percent decrease in profits over the product's life. In addition to offering a competitive price, however, suppliers must meet minimum defect levels, provide predictable and reliable delivery, and shorten product design time. The challenge for companies is to establish fewer but more collaborative purchaser-supplier relationships to satisfy the demands of customers.

In the 1990s manufacturers must market to a true *global economy;* no one country is capable of controlling the marketplace. Most manufacturers recognize the good and bad news associated with a global marketplace. Global customers offer an expanded base for marketing products; however, competition is increasing from every corner of the world. The challenges for US companies are to provide products that meet world-class standards and to market the goods and services to the global economy just as effectively as international competition.

Another external challenge presented to manufacturers is the *cost of money.* With interest rates high on a relative basis, the company that uses capital resources effectively will be the long-term winner. The cost to introduce and sustain a product in the marketplace has reached staggering proportions. Manufacturing methods that reduce the cost of doing business, such as inventory reduction, become the tools for survival.

The last and most difficult external challenge for manufacturing is provided by the *customers.* Today's sophisticated shopper buys on the basis of quality, service, cost, performance, and individual preference. If a manufacturer cannot meet all the needs of the customer, then another supplier is usually ready to step in for the sale.

Companies must respond to the external challenges (niche market entrants, traditional competition, suppliers, global economy, the cost of money, and customers) in two steps:

1. Recognize that these challenges exist and admit that the problems they create must be solved, because they will never go away.
2. Develop a manufacturing strategy to minimize the negative impact of the external challenges on the success of the business.

Internal Challenge

The internal challenge is to develop a manufacturing strategy that satisfies the following definition:

A manufacturing strategy is a plan or process that forces agreement between the market-driven corporate goals and the production capability of the company.

The goal of the manufacturing strategy is to increase product market share. However, to win orders and increase market share, the enterprise must be competitive in price, quality, design lead time, delivery reliability, flexibility, customer service, innovation, and any other factor that customers use to make a purchasing decision. These competitive qualities were identified and called *order-winning criteria* by Terry Hill, a British manufacturing engineer. Hill defined order-winning criteria *as the minimum level of operational capability required to get an order.* For example, if quality is an order-winning criterion, then the manufacturing unit must provide that feature for a product to sell. When design innovation is the criterion, then the design group is responsible. Order-winning criteria such as after-sales service, delivery reliability, and flexible financial policies demand action from areas outside of manufacturing. The identified order-winning criteria for a product give every unit or group in the enterprise its marching orders for success.

Hill also defined a second term, *order-qualifying criteria.* It is important to make the distinction between the criteria that win product orders and those that qualify a product to enter the race. An example best describes this difference. In the 1970s the order-winning criterion for the color television market in the United Kingdom (UK) was price. The Japanese entered the market with products that were competitive in price but far superior to the UK products from the standpoint of quality and reliability. Demand for imported televisions grew because of higher quality and reliability; quality became the new criterion on which purchases were made. UK manufacturers lowered the price on domestic sets to regain market share; however, the price could not drop low enough for the UK companies to sell sets, because customers were now buying because of quality, not price. Quality and reliability were the new order-winning criteria for selling televisions in the UK market. By the 1980s, UK companies matched the quality and reliability standards set by the Japanese; as a result, the order-winning criterion reverted back to *price*, and *quality* became the order-qualifying criterion.

The enterprise must meet the order-qualifying criteria to get into the market or to maintain its present market share. After the qualifying criteria are satisfied, the enterprise turns its attention to the criteria on which orders are won. Market share is increased when the order-winning criteria are understood and executed better than the competition. Therefore, in the manufacturing model developed by Hill, the enterprise must take these actions:

■ Analyze every product and agree on the order-qualifying and order-winning criteria for the current market conditions for every product.
■ For every product, project the order-winning criteria in the market in the future.
■ Determine the *fit* between the criteria necessary to succeed in the marketplace and the current capability in manufacturing.
■ Change or modify either the marketing goals or the manufacturing process choices and infrastructure to force internal consistency.

Meeting the Internal Challenge

A review of the order-winning criteria indicates that manufacturing has the responsibility to satisfy the criteria most of the time. For manufacturing to step up to this responsibility, automated systems using robotics are required to improve shop floor standards. A table of manufacturing standards compiled by the consulting company Coopers and Lybrand is given in Figure 1–3 where the US average is compared with the world-class standard. How well a company compares with the best in the world is an indication of how ready that company is to meet the criteria necessary to win orders. The values used for the US averages are open for debate, because they change with industry group and geographic area studied. Nevertheless, it is important for every company to determine how well it is performing compared with other industries in its market sector.

The following sections describe each standard and indicate some of the order-winning criteria that are affected.

	Manufacturing standards	
Attribute	*World-class standard*	*US average*
Setup time		
System	< 30 minutes	24 hours
Cell	< 1 minute	
Quality		
Captured	1500 ppm	3–5%
Warranty	300 ppm	2–5%
Cost of quality	3–5%	15–25%
Manufacturing/total space	> 50%	25–35%
Inventory		
Product velocity	> 100 turns	2–4 turns
Material residence time	3 days	3 months
Flexibility	270 parts	25 parts
Distance	300 feet	> 1 mile
Uptime	95%	65–75%

Figure 1–3 Internal Challenges. (*Source:* Rehg, James A., *Computer Integrated Manufacturing.* Englewood Cliffs, NJ: Prentice Hall; 1994, p. 12)

Setup Time.   Setup time is the length of time required to prepare a machine for production. Long setup time causes large production lot sizes and results in higher costs due to large inventories. Also, with more time allocated to setting up the machine, less time is spent in production on the machine, so more equipment and facilities are required and capital expenditures increase. The flexibility provided by robots reduces the setup time in many flexible manufacturing cells. The order-winning criteria affected include *price, delivery speed,* and *flexibility.*

Quality.   Quality is expressed as a percentage of total sales; and the three measurements that are part of it are listed in Figure 1–3. *Captured quality* is a measure of defects that are found before the product is shipped to the customer. *Warranty quality* represents defective parts discovered after the product is shipped to the customer. *Cost of quality* represents the total cost to the enterprise in percent of sales. Robots offer many quality advantages; for example, GM uses robots to put down the front windshield sealant and install the glass (Figure 1–4). The high degree of placement accuracy and repeatable performance of the robot ensure that no water

Three-roll wrist

Figure 1–4   Robot Installation of Windshield After Sealer is Put Down by Robot Automation. (*Source:* Rehg, James A., *Computer Integrated Manufacturing.* Englewood Cliffs, NJ: Prentice Hall; 1994, p. 343)

leaks will occur. In this case, the robot was selected over a human operator because of the quality needed. The order-winning criteria affected by this standard are *quality* and *price*.

Manufacturing Space Ratio.   Manufacturing space ratio is a measure of how efficiently manufacturing space is utilized. The total footprint of the process machines plus the area of the assembly workstations is divided by the total area occupied by manufacturing. When this ratio is large, the production operation is more efficient. The order-winning criterion affected is *price*.

Inventory. Inventory includes raw materials, partially completed parts, sub-assemblies, and finished goods stored by the manufacturer. Inventory is not necessarily evil; after all, some inventory is necessary to reduce disturbances in manufacturing and in fulfilling customer orders. The goal is to continuously reduce the inventory level while maintaining continuous and smooth production plus on-time and complete order shipments to the customer. Inventory levels are measured in terms of *inventory turns*, or *velocity*, and in *residence time*. The number of inventory turns for a product is equal to the annual cost of goods sold divided by the average inventory value. The residence time is the average number of days that a part or raw material item spends in production. The goal is reduced inventory; therefore, in terms of the measurement standard, inventory turns need to rise while residence time must fall. If this standard is performed well, parts and raw materials spend less time in the plant. As a result, the order-winning criteria that benefit are usually *price, quality, delivery speed,* and *delivery reliability.*

Flexibility. *Flexibility* is a measure of the number of different parts that can be produced on the same machine. Improved efficiency in this standard results from good part design, innovative fixture design, and selection of the optimum production machines. The robot supports this standard well, as the reprogrammable robot is one of the most flexible machines in manufacturing. Excellence in this standard primarily aids the *price* order-winning criterion.

Distance. Distance measures the total linear feet that a part travels through the plant from raw material in receiving to finished products in shipping. A high value for this standard indicates more cost for the product and decreased quality as a result of handling. Keeping the value of this standard low helps the *price* and *quality* order-winning criteria.

Uptime. *Uptime* is the percentage of time a machine is producing to specifications compared with the total time production can be scheduled. An increase in uptime means that less equipment is required for the same level of production. The total hours of operation without machine failures has increased fourfold over the past 10 years. Improvement in this standard helps the *price* order-winning criterion.

1-5 THE PROBLEM AND A SOLUTION

We have seen that the problem facing the enterprise is twofold: (1) the external challenges (niche market entrants, traditional competition, suppliers, global economy, cost of money, and customers) are present and will not go away; (2) customer order-winning and order-qualifying criteria drive the market. The enterprise must develop a manufacturing strategy to win orders based on the criteria present in the marketplace. In other words, the enterprise must change.

One way for manufacturers to change is to improve their performance on the following six worldwide standards:

1. Design and manufacturing lead time by product
2. Inventory turns by product
3. Setup times on production equipment
4. Output/productivity by product per employee
5. Total quality and level of rework
6. Number of suggestions by product for improvements per day per employee

The move to robot automation in the 1970s was driven by labor cost; companies had to become more productive. The emergence of global markets in the 1980s has moved the focus beyond productivity to satisfying the order-winning criteria of customers who can shop from around the globe. This new class of customers requires a *quality product, wide product selection, frequent product improvements*, and *new models* on a regular basis. Satisfying the six enterprise standards is necessary to meet the demands of customers.

Manufacturers are using a process called *CIM* (pronounced *sim*), computer-integrated manufacturing, to change their operational philosophy. The CIM process—first defined by Joseph Harrington, Jr., in 1973—guides manufacturers through the major changes required in the organization and helps identify the automation hardware and software necessary to meet the order-winning criteria for their products. CIM systems, supported by robotics, can produce a competitive product consistent with customer expectations.

1-6 DEFINITION OF ROBOTICS AND COMPUTER-INTEGRATED MANUFACTURING

Many single-purpose machines, often called hard automation, have some features that make them look like robots. Without some definition, it would be difficult to sort out industrial robots from the millions of automated machines so that they can be studied. The International Standards Organization (ISO) defines an industrial robot in standard ISO/TR/8373-2.3 as follows:

> *A robot is an automatically controlled, reprogrammable, multipurpose, manipulative machine with several reprogrammable axes, which may be either fixed in place or mobile for use in industrial automation applications.*

The key words are *reprogrammable* and *multipurpose,* because most single-purpose machines do not meet these two requirements. Reprogrammable implies two elements: (1) the robot's motion is controlled by a written program, and (2) the program can be modified to change significantly the motion of the robot arm. Programming flexibility is demonstrated, for example, when the pickup point for a randomly placed part is located by a vision system's camera, and the program is changed while the robot arm is moving to the part. Multipurpose means a robot must be able to perform many different functions, depending on the program and tooling currently in use. For example, in one company a robot could be tooled and programmed to do welding, and in a second company the same type of robot could be used to stack boxes on pallets.

The Computer and Automation Systems Association (CASA) of the Society of Manufacturing Engineers (SME) defines CIM as follows:

> *The integration of the total manufacturing enterprise through the use of integrated systems and data communications coupled with new managerial philosophies that improve organizational and personnel efficiency.*

CIM describes a new approach to manufacturing, management, and corporate operation. Although CIM systems can include many advanced manufacturing technologies such as robotics, computer numerical control (CNC), computer-aided design (CAD), computer-aided manufacturing (CAM), and just-in-time production (JIT), it goes beyond these technologies. CIM is a new way to do business that includes a commitment to total enterprise quality, continuous improvement, customer satisfaction, use of a *single* computer database for all product information with *every* department participating, removal of communication barriers among all departments, and the integration of enterprise resources. The CIM concept was

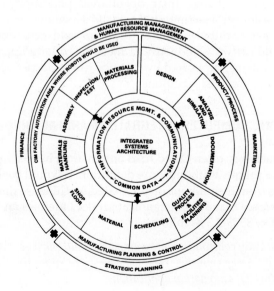

Figure 1–5   CASA/SME Wheel.

modeled (Figure 1–5) by CASA/SME as a wheel with the following functions: (1) general business management (outside ring of management functions), (2) product and process definition, (3) manufacturing planning and control, (4) factory automation, and (5) information resource management. Many industries use robots in all of the factory automation units in Figure 1–5 (material handling, assembly, inspection and testing, and material processing) and have integrated them into the enterprise CIM system. As a result, the study of robotics must address the techniques used to interface robots to the CIM system.

1-7 MANUFACTURING SYSTEM CLASSIFICATION

In many early implementations, the robot was simply dropped into the production area to replace the human operator. Little effort or planning was done to integrate the robot into the manufacturing system, because the robot was viewed as just a human replacement. Now industry understands that the robot is just one part of the total manufacturing system, and the robot must be integrated into the system along with the other automation hardware and process machines. As a result, an understanding of the manufacturing system is critical for selection of the correct robot model and supporting hardware. The study of manufacturing systems starts with a classification of manufacturing based on how products are produced.

The classification process divides all production operations into the following five groups: *project, job shop, repetitive, line,* and *continuous.* Overlap cannot be avoided between some of the categories, and most manufacturers use two or more of the manufacturing systems in the production of an entire product line. Classification of companies into these groups requires a detailed analysis and evaluation of the production operations. As a result of the classification process, the activity in the factory automation area of the CIM wheel (Figure 1–5) can be better understood. If the manufacturing system requirements are well defined, then success in the selection and integration of the robot hardware and software is assured. Distinguishing characteristics of each classification category are described next.

Project

The most distinguishing characteristics of this category is that products are complex, and production quantities are often just one unit. Products coming from a project-type manufacturing system include oil refineries, large office buildings, cruise ships, and large aircraft. In each case individual products may be similar but they usually are not identical. The layout of the production area for this manufacturing system is called *fixed position* (Figure 1–6a). Because of their size and weight, products like ships and large aircraft remain in one location and the production equipment and parts are moved to them. Few robot applications are found in project-type manufacturing because the production and assembly equipment must be mobile.

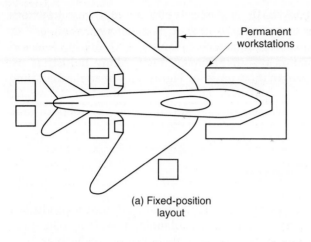

(a) Fixed-position
layout

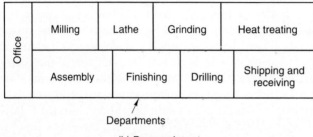

(b) Process layout

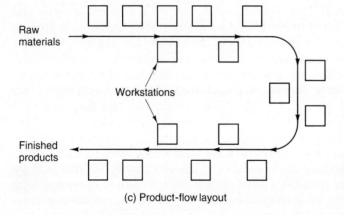

(c) Product-flow layout

Figure 1–6 Types of plant layout:
(a) fixed-position layout; (b) process
layout; (c) product-flow layout.
(*Source:* Rehg, James A., *Computer
Integrated Manufacturing.* Englewood
Cliffs, NJ: Prentice Hall; 1994, p. 27)

Job Shop

Job shop production quantities, called *lot sizes,* are small as are the part size
and weight. As a result, the parts are moved or routed between fixed production
work cells for manufacturing processing. The classic machine shop with lathes,
mills, grinders, and drill presses is the example most often used for the job shop
classification. The production equipment layout for the job shop, Figure 1–6b, is

frequently called *job shop layout* or *process layout.* Other distinguishing features include less than 20 percent repeat production on the same part, noncomplex products, and intensive movement of the products between machines and departments. The opportunity for robot applications in the job shop are present but limited by the high variation in parts and products.

Repetitive

The repetitive-type manufacturing system has the following unique characteristics: orders for repeat business approaching 100 percent, contracts with customers for multiple years, moderately high product volume with production quantities varying over a large range, and little variation in the routings of parts between production machines. The plant layout could be either the process layout in Figure 1–6b or be more like the product-flow layout in Figure 1–6c. For example, a supplier to Ford uses the repetitive-type manufacturing system to make three different water pump models with a combined volume of 10,000 pumps per week. A contract for 3 years on a product like this is very common. The production machines in this type of manufacturing system are usually special-purpose, automated systems with robots frequently integrated into the process.

Line

The line-type manufacturing system has several distinguishing characteristics: the delivery time required by the customer is often shorter than the total time it takes to build the product, the product has many different options or models, and an inventory of subassemblies is normally present. Car and truck assembly is an example of this category. Manufacturing facilities in this group usually use the product-flow (Figure 1–6c) type of plant layout. Robots are frequently used to perform assembly tasks as the product moves through the line (see Figure 1–4).

Continuous

Continuous manufacturing systems have the following characteristics: the time required for manufacturing the product is longer than the time the customer is willing to wait, product demand is predictable, product inventory is held, high production volume is present, and products have few options. This type of production system always uses the product-flow type of plant layout (Figure 1–6c) with the production line limited to one or just a few different products. The production of nylon carpet yarn is an example of this type of industry; chemical compounds enter on one end of the system and a finished product, nylon filament thread, flows out of the other end. Production of electrical components (for example, switches for the automotive industry) also involves this type of manufacturing system. The linear flow manufacturing system in Figure 1–7 illustrates high-speed line assembly. Robots used in these applications are capable of high-speed and high-volume operation.

The chart in Figure 1–8 offers comparative data on the characteristics of the manufacturing systems just described. Take some time and study the chart. We will consult this chart later to see how the characteristics of the manufacturing systems dictate the type of robot automation most frequently used.

Figure 1–7 High-Speed, In-line, Fixed Automation System for Small Parts. (Courtesy of METO-FER, Corp., Pittsburgh, PA)

	Project	Job shop	Repetitive	Line	Continuous
Process speed	Varies	Slow	Moderate	Fast	Very fast
Labor content	High	High	Medium	Low	Very low
Labor skill level	High	High	Moderate	Low	Varies
Order quantity	Very small	Low	Varies	High	Very high
Unit quantity cost	Very large	Large	Moderate	Low	Very low
Routing variations	Very high	High	None	Low	Very low
Product options	Low	Low	None	Very high	Very low
Design component	Very large	Large	Very small	Moderate	Small

Figure 1–8 Manufacturing System Characteristics.

The number of manufacturing classification categories is not critical; however, the classification system is important in understanding what makes automotive production different from nylon production. A specific robot model may satisfy the automation requirements of just one manufacturing classification category. So a successful robot implementation requires complete knowledge of the manufacturing system to be automated, and classification is the first step.

1-8 ROBOT SYSTEMS

Integrating robots into the repetitive, line, and continuous manufacturing systems requires knowledge of both the manufacturing and robot systems. Therefore, the definition of the robot system is the logical starting point. The Robotic Industries Association defines a robot as follows:

> An industrial robot system includes the robot(s) (hardware and software) consisting of the manipulator, power supply, and controller; the end-effector(s); any equipment, devices, and sensors with which the robot is directly interfacing, any equipment, devices and sensors required for the robot to perform its task; and any communications interface that is operating and monitoring the robot, equipment, and sensors.

As the definition indicates, a robot system is more than the hardware; it includes any devices interfaced to the robot for control of the work cell. In the chapters that follow, the larger robot system detailed in the definition will be studied. At this time, however, let us consider the *basic robot system.*

Basic System

A basic robot system is illustrated in the block diagram in Figure 1–9. The system includes a mechanical arm to which the end-of-arm tooling is mounted, a computer controller with attached teaching device, work-cell interface, and program storage device. In addition, a source for pneumatic or hydraulic power is a part of the basic system. The work-cell controller and other external devices that are a part of the manufacturing system connect to the robot through the robot work-cell interface.

Mechanical Arm

The arm is a mechanical device driven by electric-drive motors, pneumatic devices, or hydraulic actuators. The basic drive elements will be either linear or rotary actuators. The combination of motions included in the arm determine the type of arm geometry that is present. The basic geometries include rectangular, cylindrical, spherical, and jointed-spherical. Figure 1–10 shows a jointed-spherical mechanical arm with all the motions indicated. Note that all six arm motions are rotational, and that five of the six are produced by rotational actuators. The elbow extension is the only rotary motion produced by a linear actuator.

The six motions are divided into two groups. The first group includes the arm sweep, shoulder swivel, and elbow extension and is called the *position*

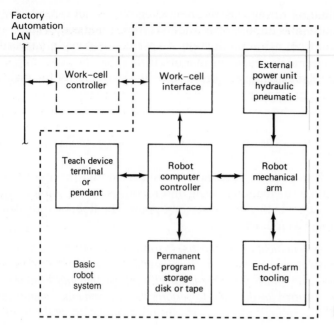

Figure 1–9 Basic Robot System.

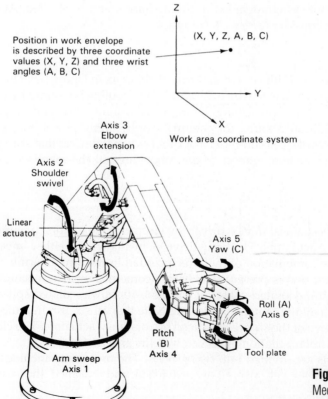

Position in work envelope is described by three coordinate values (X, Y, Z) and three wrist angles (A, B, C)

(X, Y, Z, A, B, C)

Work area coordinate system

Axis 3
Elbow extension

Axis 2
Shoulder swivel

Linear actuator

Axis 5
Yaw (C)

Roll (A)
Axis 6

Pitch (B)
Axis 4

Tool plate

Arm sweep
Axis 1

Figure 1–10 Jointed-Spherical Mechanical Arm. (Courtesy of Cincinnati Milacron Corp.)

motions. With a combination of these three motions, the arm can move to any required position within the work area. The second group of motions, associated with the wrist at the end of the arm, includes the pitch, yaw, and roll. A combination of these three motions, called *orientation*, permits the wrist to orient the tool plate and tool with respect to the work.

The T3-566 arm in Figure 1–11 is late 1970s technology, but it is an excellent model to study because all the drive elements are visible. A complete description of the arm geometries, drive systems, and controller operation is presented in Chapter 2.

Production Tooling

The robot arm alone has no production capability, but the robot arm interfaced to production tooling becomes an effective production system. The tooling to perform the work task is attached to the tool plate at the end of the arm. The tool plate, which is usually a part of the wrist, is identified in Figure 1–10.

The tooling is frequently identified by several names. The term used to describe tooling in general is *end-of-arm tooling* or *end effector.* If the tooling is an

"F" for elbow

"B" for shoulder

"C" for pitch

"D" for yaw

"A" in base

"E" for roll

Figure 1–11   Cincinnati Milacron T3-566 Mechanical Arm.

Figure 1–12 Fibro Manta Gripper.

open-and-close mechanism to grasp parts, it is referred to as a *gripper* (Figure 1–12). In this text the terms *gripper* and *end-of-arm tooling* will both be used to refer to the tool on the robot arm. A complete discussion of robot end-of-arm tooling is presented in Chapter 4.

External Power Source

The external power required to operate a robot system includes sources to drive the arm motion (*electrical, hydraulic,* or *pneumatic*) and electricity for the electronic controller. Because most grippers are activated by compressed air, a source of compressed air is required for most systems. Large robot arms using hydraulic actuators for motion require a hydraulic power source, and in some cases a compressed air source will also be necessary for the gripper. All electric-drive arms require only electrical power for motion, but many need compressed air for tooling. Hydraulic robots use a pump and tank assembly (Figure 1–13) to generate the 2300-psi fluid pressure necessary to operate the arm.

Robot Controller

Of all the building blocks of a robot system, the controller is the most complex. It is also the unit with the greatest degree of variation from one manufacturer to the next. Figure 1–14 is a basic block diagram of a typical controller used on electric robots.

The controller, basically a special-purpose computer, has all the elements commonly found in computers, such as a central processing unit (CPU), memory, and input and output devices. Most controllers have a network of CPUs, usually standard microprocessors, each having a different responsibility within the system (Figure 1–14). The distributed microcomputer network in the controller has the primary responsibility for controlling the robot arm and the work cell in which it is operating. The controller receives feedback from the arm on joint position and velocity and responds with outputs to the servo drives to change the current posi-

Figure 1–13 Power Unit for Hydraulic Robot.

tion or velocity based on the program stored in memory. The controller can also communicate with external devices through the *input/output interface* (Figure 1–14). One external device, an industrial cassette recorder used for program storage, is pictured in Figure 1–15. In addition to the tape storage capability, most current controllers use the standard DOS or Windows program format with $5\frac{1}{4}$- or $3\frac{1}{2}$-inch disk drives to save programs permanently. The operation of controllers is discussed in detail in Chapter's 2 and 7.

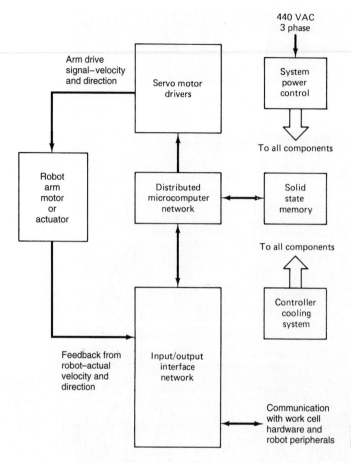

440 VAC
3 phase

Arm drive
signal–velocity
and direction

Servo motor
drivers

System
power
control

To all components

Robot
arm
motor
or
actuator

Distributed
microcomputer
network

Solid
state
memory

To all components

Controller
cooling
system

Feedback from
robot–actual
velocity and
direction

Input/output
interface
network

Communication
with work cell
hardware and
robot peripherals

Figure 1–14 Robot Controller
Block Diagram.

Teach Stations

Teach stations on robots may consist of teach pendants, teach terminals, or
controller front panels. The teach terminal in Figure 1–16 is an example of a handheld terminal with a flat screen, multiline, touch-sensitive display. Some robots
permit a combination of programming devices to be used in programming the
robot system, whereas others provide only one system programming unit. In addition, some robots use IBM PC or compatible microcomputers as programming stations. Teach stations support three activities: (1) robot *power-up* and preparation for
programming; (2) *entry* and *editing* of programs; and (3) *execution* of programs in
the work cell. The development of a program includes keying in or selecting menu
commands in the controller language, moving the robot arm to the desired position in the work cell, and recording the position in memory. Chapter 8 includes a
detailed analysis of the programming process.

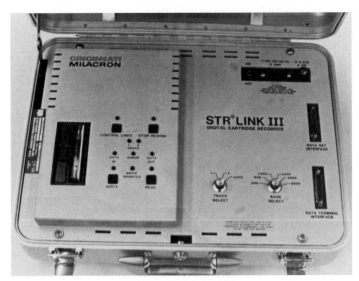

Figure 1–15 Permanent Program Storage.

Figure 1–16 Touch-Sensitive Screen Teach Pendant. (Courtesy of Sankyo Seiki (America) Inc.)

1-9 SOME BASIC TERMS

With the introduction of every new technology there follows a new vocabulary that must be learned to be literate in that field. Robotics is no exception. Most of the robotics vocabulary can be learned as required, but a few terms are prerequisites. Specifications for an AdeptOne robot (Figure 2–18) from Adept Technology, Inc., are given in Figures 1–17 and 1–18; compare the values with those given in the following definitions.

AdeptOne Robot

Product Specifications

Load

Payload:	20 lbs (9.09 kg)
Downward force:	40 lbs (18.2 kg)
Joint 4 inertia	
standard:	96 lb-in^2 (280 kg-cm^2)
maximum:	1000 lb-in^2 (2900 kg-cm^2)

Joint Motion

Joint 1:	300°
Joint 2:	294°
Joint 3	
standard:	7.7"
optional:	11.6"
Joint 4:	554°

Positioning (X,Y)

Resolution:	0.0006"
Repeatability:	0.001"
Accuracy††:	0.003"

Robot Brakes

Joints 1, 2, and 3 use fail-safe mechanical brakes

End-of-Arm User Connections

User solenoid valves:	2
Additional 12VDC solenoid drivers:	2
User signal lines:	10

Robot-to-Controller Cable 15'

Design Life 42,000 hours

Calibration Method

Uses local calibration indices

Robot Weight 400 lbs

Cycle Time

Maximum velocity (no load): 30 ft/sec

The robot tool performs a continuous-path motion from location "a" to location "b" and back to "a". The path consists of all straight-line segments as shown:

Payload	Cycle Time
1 lb	0.9 sec
13 lbs	1.3 sec
20 lbs	1.7 sec

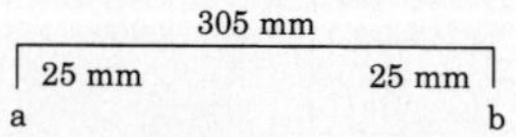

305 mm
25 mm
25 mm
a
b

Note: These cycle times apply to a standard AdeptOne. With the optional Adept HyperDrive performance enhancements, tool-tip velocity and cycle times improve. Refer to the Adept HyperDrive datasheet for details.

† *maximum possible using "J4 Tuner" software and running at reduced speed.*
†† *over a 17" X 17" area using Adept's High-accuracy Positioning System (HPS).*

Options

- Adept HyperDrive
- High-accuracy Positioning System (HPS)
- Force Sensing Module
- Clean room package (Class 10)
- Fifth axis (servo pitch axis)
- 11.6" vertical stroke
- Robot-mounted camera calibration
- 25' robot-to-controller cable set
- Conveyor-tracking hardware

System Requirements

Controller

Compatible with Adept A-series or S-series MC and CC controllers.

Power

The Adept controller supplies all necessary power and control.

Air

80 psi minimum, 120 psi maximum, 1 CFM, clean dry air (excluding end effector requirements).

Environment

41 – 122°F (5 – 50°C), 5 – 90% relative humidity (noncondensing).

Product specifications are subject to change without notice
Adept, AdeptOne, Adept MC, Adept CC, and HyperDrive are trademarks of Adept Technology, Inc.
Revision date 9/92.

Figure 1–17 AdeptOne Robot Specifications. (Courtesy of Adept Technology, Inc.)

AdeptOne Manipulator Dimensions

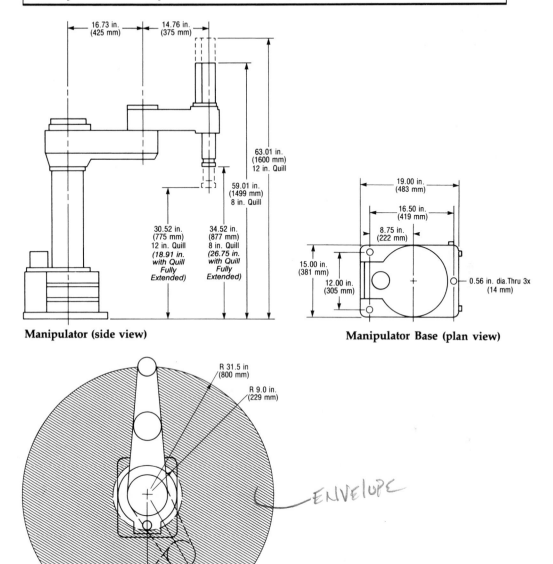

Manipulator (side view)

Manipulator Base (plan view)

Manipulator Envelope

Figure 1–18 AdeptOne Robot Work Envelope and Arm Dimensions. (Courtesy of Adept Technology, Inc.)

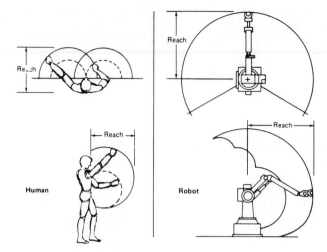

Figure 1–19 Human and Robot Work Envelope. (Courtesy of TPC Training Systems)

Work Envelope. The space in which the robot gripper can move with no limitations in travel other than those imposed by the joints. Figure 1–19 shows a work envelope or work volume for a spherical geometry robot in contrast to that of a human worker.

Axis Numbering. The following axis numbering convention is used to describe the axes of motion of a robot (see Figure 1–10).

1. Start at the robot mounting plate (base).
2. Label the first axis of motion encountered as axis 1.
3. Progress from the base to the end-effector numbering each axis encountered successively.

Position Axes. The position axes for the robot in Figure 1–10 are labeled 1, 2, and 3. An ancillary device often used to provide an additional axis, *linear movement* of the entire robot along the floor, is shown in Figure 2–9. The position axes, defined for each robot geometry, move the tooling to the point where work must be performed.

Orientation Axes. The orientation axes, labeled 4, 5, and 6 in Figure 1–10, orient the tooling so that it is aligned with work. *Roll, pitch,* and *yaw* are illustrated in Figure 1–20 and compared with those of the human wrist (Figure 1–20a). Three techniques are used in robotics to duplicate the human movement. The first configuration (Figure 1–20b) was described earlier. The second (Figure 1–20c) uses a roll 1, pitch/yaw, and roll 2 configuration. Note that the center axis produces yaw in the vertical position and pitch in the horizontal position. The *three-roll wrist,* the third type, is shown in Figure 1–4.

Degree of Freedom. Every joint or movable axis on the arm is a degree of freedom. A machine with six movable joints, such as the one in Figure 1–10, is a robot

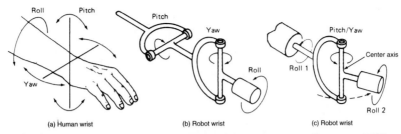

Figure 1–20 Human and Robot Wrist with Orientation Axes. (Courtesy of TPC Training Systems)

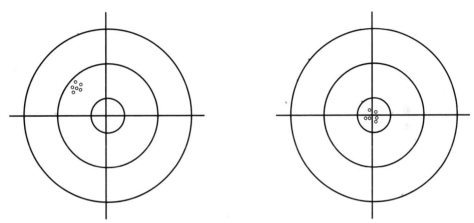

Figure 1–21 Rifle Targets: (a) Repeatable and (b) Accurate and Repeatable.

with 6 *degrees of freedom* or *axes*. Orientation of the tool by the wrist involves a maximum of 3 degrees of freedom, and up to 4 degrees of freedom are used for positioning within the work envelope. A range or 4 to 7 degrees of freedom is typical for industrial robots.

Coordinate Systems. All points programmed in the work cell are identified by a *base coordinate system* that consists of three translation coordinates—X, Y, and Z—and three rotational coordinates—A, B, and C. The number of work-cell coordinates required to define a programmed point is determined by the number of degrees of freedom present on the robot. Figure 1–10 shows the coordinate system frequently used by robots with 6 degrees of freedom.

Accuracy. Accuracy is best explained by an example from target shooting with a rifle. The targets in Figure 1–21 indicate the results when two different rifles are fired at the center of the targets. The rifle producing the results in Figure 1–21a is not accurate because all shots missed the center; however, the rifle is repeatable because all shots hit the target in the same area. The second rifle was both accurate

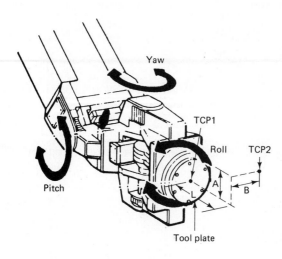

Figure 1–22 Tool Center Point. (Courtesy of Cincinnati Milacron Corp.)

and repeatable, because it placed all shots (Figure 1–21b) in a close group in the center of the target.

In robotics, accuracy is the degree to which a robot arm is able to move to a specific *translation* or *position* point in the work cell when the point coordinates are entered from an off-line programming station, calculated inside the program, received from a vision system, or generated in a work-cell simulator. For example, a vision system could specify that the robot tooling should move to a point described as X = 50.00 inches, Y = 45.300 inches, Z = 10.01 inches (tooling position), A = 0.00 degrees, B = 90.00 degrees, and C = 100.00 degrees (tooling orientation). The accuracy specification describes how close the gripper will be to the point described by the vision system. Robot accuracy is usually one or two magnitudes worse than the arm's repeatability.

Repeatability. Repeatability is the degree to which a robot system is able to return to a specific position point in the work cell. Frequently, a robot is taught the required gripper location by moving the arm to the location with the teach pendant and then pushing the *program point* button. The repeatability specification indicates how well the robot arm can return to the taught point on each cycle of the program execution. The best repeatability on assembly robots is ± 0.0005 inches.

Offset. The *point of action* for the tool mounted to the robot tool plate is called the *offset* or *tool center point* (TCP). With an offset of 0, 0, 0 for L, A, and B, respectively, the TCP is located at point TCP1 in Figure 1–22. With L = 10.00 inches, A = 5.00 inches, and B = 4.00 inches, the TCP is located at TCP2. When a tool is mounted to the tool plate, the distance from the TCP1 to the action point on the tool is included in the robot program. For example, the welding torch in Figure 1–23 has a TCP of L = 14.00 inches, A = −4.00 inches, and B = 0 inches. The robot will control the motion at the welding electrode as the arm moves through the programmed

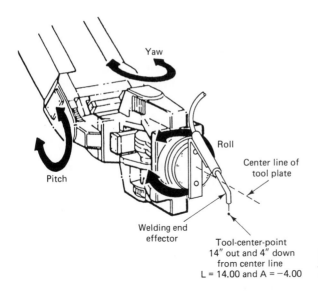

Yaw

Roll

Center line of
tool plate

Pitch

Welding end
effector

Tool-center-point
14″ out and 4″ down
from center line
L = 14.00 and A = −4.00

Figure 1–23 Robot with Welding Tooling.

points. In some robot systems the offset is a part of the programming language. In the Seiko DARL language, for example, the command is **DEF TL** $<1-9>$ X_1, Y_1, X_2, Y_2, Z_1. Up to nine define tool statements can be used to specify the X, Y, and Z coordinates for the TCP.

Velocity. The rate at which the robot can move each axes and the TCP under program control is a measure of the machine's velocity. Velocity is expressed in linear or angular English and metric units. Arm velocity between programmed points is much higher than the average velocity over a number of programmed points because of the deceleration and acceleration of the TCP as it passes through a programmed point. Direct-drive robots like the AdeptOne have excellent tool velocity; note the velocity specification in Figure 1–17.

Payload. The *rated* payload is the mass that the robot is designed to manipulate under the manufacturer's specified performance conditions of speed and acceleration-deceleration over the entire work envelope. The center of gravity of the payload must be at offsets or locations specified by the manufacturer. The *maximum* payload is the maximum mass that the robot can manipulate at a specified speed, acceleration-deceleration, center of gravity location (offset or location), and repeatability under continuous operation over a specified work envelope.

The payload or load capacity of a robot (Figure 1–17) often determines if the machine is suited for a specific task. The payload is the total combined weight of the gripper or end-of-arm tooling and the part to be moved (payload = tooling weight + part weight).

Additional terms associated with the robot hardware and software are introduced as needed throughout the text.

1-10 ROBOT STANDARDS

In 1984 the Robotic Industries Association (RIA) established the R15 Executive Committee on Robotic Standards and the Automated Imaging Association (AIA). One year later the A15 Standards Committee in the AIA was formed to deal with vision-related standards. The robot standards will follow the procedures of the American National Standards Institute (ANSI).

R15 Standards

The following standards are addressed by the R15 subcommittees:

- **R15.01 ELECTRICAL INTERFACE.** ANSI/RIA R15.01: *American National Standard for Industrial Robots and Robot Systems—Common Identification Methods for Signal and Power Carrying Conductors.*
- **R15.02 HUMAN INTERFACE.** ANSI/RIA R15.L02: *American National Standard for Industrial Robots and Robot Systems—Human Engineering Design Criteria for Robot Control Pendants.*
- **R15.03 MECHANICAL INTERFACE.** Two standards: (1) ANSI/RIA R15.03/1: *Circular Mechanical Flange Interface;* (2) ANSI/RIA R15.03/2: *Shaft Mechanical Interface.*
- **R15.04 COMMUNICATION/INFORMATION.** Companion standard to the ISO 9506 standard on Manufacturing Message Specification.
- **R15.05 PERFORMANCE.** ANSI/RIA R15.05: *American National Standard for Industrial Robots and Robot Systems—Evaluation of Point-to-Point and Static Performance Characteristics.* Additional standards on dynamic characteristics and reliability are also available.
- **R15.06 SAFETY.** ANSI/RIA R15.06: *American National Standard for Industrial Robots and Robot Systems—Safety Requirements.*
- **R15.07 SIMULATION/OFFLINE PROGRAMMING.** ANSI/RIA R15.07: *American National Standard for Industrial Robots and Robot Systems— Simulation/Off-line Programming—Terms and Notations for Characterizing Industrial Robots.*

A15 Standards

The following standards are addressed by the A15 subcommittees:

- **A15.01 SYSTEM COMMUNICATION.** Communication standards and protocols for exchange of information between machine vision systems and other devices.
- **A15.05 PERFORMANCE.** ANSI/AVA A15.05/1—1989: *Automated Vision Systems—Performance Test—Measurement of Relative Position of Target Features in Two-Dimensional Space.*
- **A15.07 TERMINOLOGY.** Terminology standards to facilitate machine vision communication including person to person, data communication, and user interfaces.

- **A15.08 SENSOR INTERFACES.** These are a series of machine vision camera interfaces standards.
- **A15.09 MARKING AND LABELING.** Marking and labeling standards for machine vision systems.

The terminology and definitions used throughout the text are consistent with the RIA and AIA standards listed previously.

1-11 SUMMARY

Automation has changed the workplace, and CIM will bring even greater change. A study of the history of robotics indicates that the technology is maturing and has a role in the CIM factory of the 1990s. Although economics and technology caused the initial move to robots in the 1970s, it is the drive to meet customer demands that fuels the move to current robot applications.

The definition of robot hardware includes two important terms: reprogrammable and multipurpose. To qualify as a robot, the machine's hardware must be capable of changing its functional characteristics by changing the program that drives it through its motions. The basic robot system includes a mechanical arm, special tooling attached to the arm at the tool plate, one or more computers in a controller, a teach station, a program storage device, and a source of pneumatic or hydraulic power.

QUESTIONS

1. Identify an early robot-like device that demonstrates the mechanical operation found in later industrial robots.
2. Who was awarded the first patent for an industrial robot, and in what year was the patent approved?
3. What were the names of the first three robot manufacturers?
4. What company was the first to drive a robot with a powerful minicomputer?
5. Describe the two major events that caused the emergence of robots in the 1970s.
6. Why didn't implementation of robots in the 1980s follow earlier predictions?
7. Describe the six major external factors that affect manufacturing, and identify the factor with the greatest influence on the products offered.
8. What steps are necessary to overcome the effects of the external challenges?
9. What are order-winning and order-qualifying criteria, and how do they affect decisions on adding manufacturing automation and robots to the shop floor?
10. Describe the seven shop-floor standards that drive efficient manufacturing, and identify those that are supported most effectively by robot technology.
11. Why has the emergence of the global marketplace forced companies to move to integrated solutions and adopt CIM-type managerial philosophies?

12. What is the CASA/SME definition of CIM?

13. What is the ISO definition of a robot?

14. What operational feature makes robots different from other hard-automation machinery?

15. What is the relationship between robotics and CIM?

16. How are current robot work-cell implementations different from the cells constructed in the 1970s?

17. How do the characteristics of the five basic manufacturing systems affect the adoption of robot automation?

18. What are the major elements of a robot system?

19. What is the difference between position and orientation arm motions?

20. What are the three names used to identify robot tooling?

21. What function does the external power source serve in the basic robot system?

22. Describe the basic elements found in robot controllers.

23. What are the three basic operations performed from robot teach stations?

24. Define repeatability and accuracy with respect to robot operation and explain how they are different.

25. Define the following robot terms: work envelope, degrees of freedom, orientation axes, position axes, tool center point, maximum operational payload, velocity, and work-cell coordinates.

PROBLEMS

1. A manufacturer plans to use an AdeptOne robot (Figures 1–17, 1–18) to assemble electric motors. If the weight of the motor armature is 6 pounds, what is the maximum allowed weight for the tooling? What is the maximum tooling weight for a 50-percent safety factor?

2. Use the robot specifications in Figures 1–17 and 1–18 and calculate the time required to move at maximum velocity over a 25-inch straight path with no load.

3. The program for the robot in problem 1 moves the TCP through 425 inches of motion with the part in the tooling and 350 inches with the gripper empty. The total program has 45 programmed points. Calculate the time for the robot to complete one cycle moving at 30 percent of maximum velocity when the part is in the gripper and 60 percent when the gripper is empty. Assume time to pass through a program point is 0.08 seconds and the maximum velocity is the same as the AdeptOne.

4. Repeat problem 1 for another robot. (Use the data sheet given to you by your instructor or one from another source.)

5. Repeat problem 2 for another robot. (Use the data sheet given to you by your instructor or one from another source.)

6. Repeat problem 3 for another robot. (Use the data sheet given to you by your instructor or one from another source.)

PROJECTS

1. Create a two-dimensional or three dimensional wire frame or solid model of the AdeptOne robot arm (Figure 1–18) using AutoCAD or any other available CAD software.

2. Use primary and/or secondary research to gather data on five manufacturers of different products in your city, region, or state that use automated manufacturing systems with robots integrated into the production process. The data should include the company name, address, and phone number, products manufactured, and type of manufacturing system used.

3. Using available resources, determine the current order-winning and order-qualifying criteria for products identified in project 2.

4. Prepare a report on the concept of order-winning and order-qualifying criteria described in the book, *Manufacturing Strategies,* by Hill.

5. Prepare a report on the definition of CIM as described by Harrington in his book *Computer Integrated Manufacturing.* Include a comparison with the SME definition provided in this chapter.

Robot Classification

2-1 INTRODUCTION

Work today tends to put everything into categories, groups, and classifications. The robot industry is no exception. In this chapter the five machine classification groups (*arm geometry, power sources, applications, control techniques,* and *path control*) used by the robot industry are explained so that we can develop a working knowledge of current robots. In addition, the classification technique used by the International Standards Organization is described. This chapter also provides an opportunity to compare and contrast the primary system components that make each vendor's hardware unique.

2-2 ROBOT ARM GEOMETRY

In general, the basic mechanical configurations of the robot manipulator are categorized as *Cartesian, cylindrical, spherical,* and *articulated*. The Cartesian is divided into traverse axes and gantry, and the articulated is divided into horizontal and vertical. A description of these configurations follows.

Cartesian Geometry

A robot with a *Cartesian* geometry is able to move its gripper to any position within the cube or rectangle defined as its work envelope. Two configurations form this geometry, *traverse-*and *gantry-*type machines. Figure 2–1 is an example of the

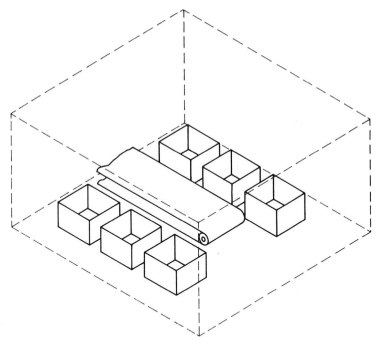

Figure 2–1 Rectangular Work Volume.

rectangular work volume that a robot must use to load a conveyor from supply bins. A Cartesian gantry robot, like that illustrated in Figure 2–2 and pictured in Figure 2–3, could load the supply bins. In Figure 2–2, the 3 degrees of freedom for positioning that movement in the X, Y, and Z directions are indicated by arrows on axes 1, 2, and 3; the 3 degrees of freedom for orientations A, B, and C of the tool mounting plate on the wrist are also shown by arrows. A Cartesian traverse-type robot is pictured in Figures 2–4 and 2–5. A common characteristic of the traverse style is the mounting of positional axes on top of each other. For example, the vertical Z axis in Figure 2–4 is mounted to the horizontal X axis, and the Y axis is mounted on the Z axis.

The power for movement in the X, Y, and Z directions is provided by pneumatic linear actuators or by ball-screw drives. However, rotary actuators are used to provide the pitch, roll, and yaw turning motion at the tool mounting plate; the positioning and orientation actuators are powered by hydraulic, pneumatic, or electric sources.

Cartesian coordinate geometry has the following advantages:

- Very large work envelopes are possible because travel along the X axis can be increased easily. Systems have been developed with work envelopes more than 80 feet long.
- Overhead mounting leaves large areas of manufacturing floor space free for other uses.
- Simpler control systems can be used.

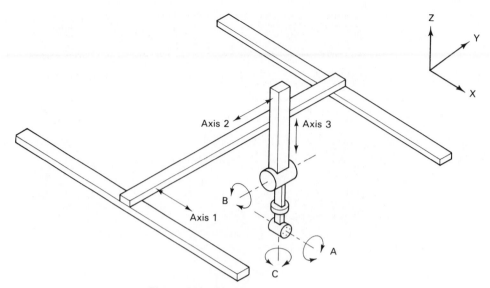

Figure 2–2 Rectilinear Geometry Robot.

Figure 2–3 Large Gantry Robot to Load and Unload Process Machines.

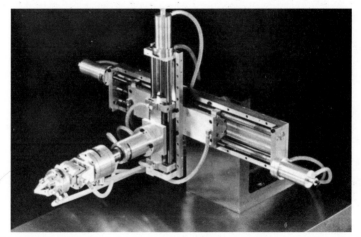

Figure 2–4 B-A-S-E Robot System from Mack Corporation. (Courtesy of Mack Corporation)

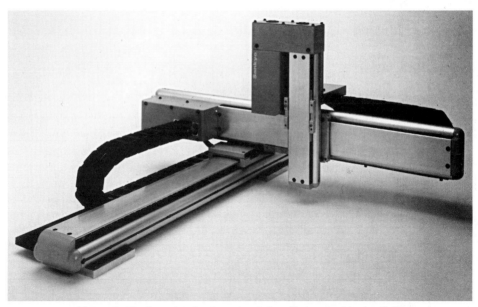

Figure 2–5 Cartesian Configuration Robot. (Courtesy of Sankyo Seiki (America) Inc.)

The disadvantages of this type of robot geometry include the following factors:

- Access to the work envelope by overhead crane or other material-handling equipment may be impaired by the robot-supporting structure.
- On some models the location of drive mechanisms and electrical control equipment overhead makes maintenance more difficult.

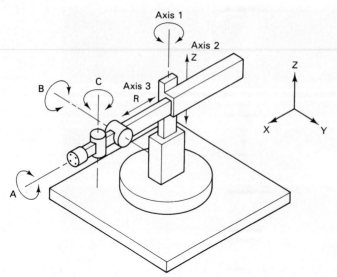

Figure 2–6 Cylindrical Geometry Robot.

The primary applications for Cartesian coordinate systems are in material handling, part handling related to machine loading and unloading, assembly of small systems, and in electronic printed circuit board assembly.

Cylindrical Geometry

A *cylindrical* geometry robot, like that illustrated in Figure 2–6, can move its gripper within a volume that is described by a cylinder. The cylindrical geometry arm is positioned in the work area by two linear movements along the Z axis and in the R direction, usually using ball-screw drives (Figure 2–7), and one angular rotation about the Z axis. In the ball-screw drive shown, a motor turns the *ball mechanism*, which has internal ball bearings riding in the threads of the screw. The screw and yoke are driven in a linear motion by rotation in the captured ball. The motions of the Z and R axes are illustrated in Figure 2–8a and b. Movement of the gripper in the cylindrical geometry work cell requires a controller that can coordinate the motion of all of the axes during a move. As a result, the controller must be more sophisticated than the controller used for Cartesian geometry systems.

The axes on cylindrical geometry robots are driven pneumatically, hydraulically, or electrically . The axis 1 rotation for the cylindrical geometry robot in Figure 2–6 is less than 360 degrees because of mechanical design limitations.

Some of the advantages of the cylindrical geometry are as follows:

- Deep horizontal reach into production machines is possible.
- The vertical structure of the machine conserves floor space.
- A very rigid structure is possible for large payloads and good repeatability.

The singular disadvantage is the limited reach to left and right because of the mechanical constraints that limit the size of the horizontal actuator. This is often

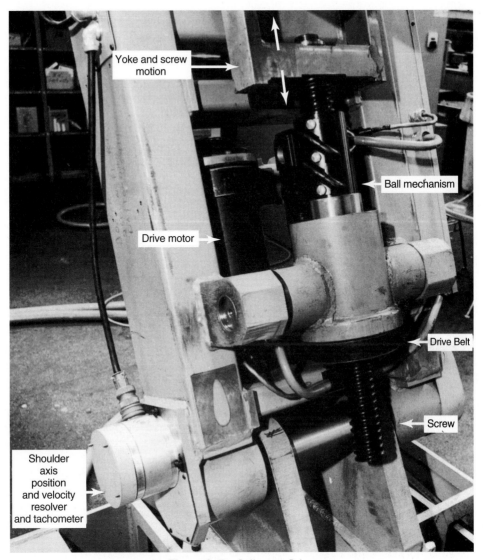

Yoke and screw
motion

Ball mechanism

Drive motor

Drive Belt

Screw

Shoulder
axis
position
and velocity
resolver
and tachometer

Figure 2–7   Ball-screw Drive.

overcome by mounting the robot on a movable platform that can be positioned anywhere along the Y coordinate. Figure 2–9 illustrates this type of linear base.

The cylindrical type of arm geometry can be used for most applications, but it is especially desirable when deep horizontal reach is necessary or when the manufacturing layout consists of machines to be serviced by the robot in a circle with a small radius. The die-cast application in Figure 2–10 is an example of this type of production layout. A small pneumatic cylindrical geometry robot is pictured in Figure 2–11.

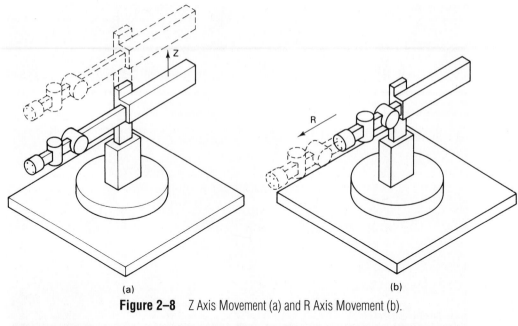

Figure 2–8 Z Axis Movement (a) and R Axis Movement (b).

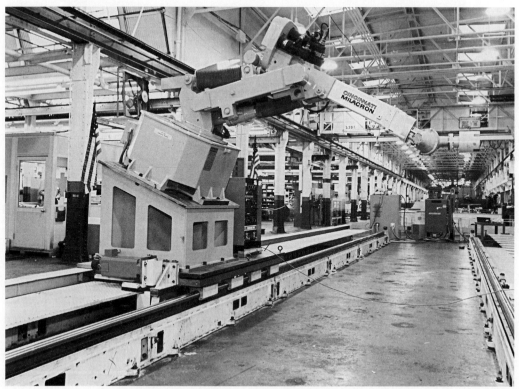

Figure 2–9 Jointed Spherical Robot Mounted to a Long Linear Axis.

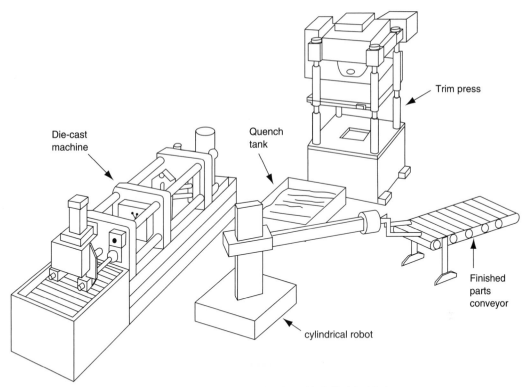

Figure 2–10 Die-cast Application with Cylindrical Robot.

Figure 2–11 Seiko Model 700 Robot.

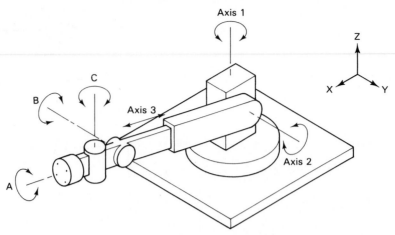

Figure 2–12   Spherical Geometry Robot.

Spherical Geometry

The *spherical* geometry arm (Figure 2–12), sometimes called *polar*, requires coordinated motion in every positioning axis for movement in the X, Y, or Z directions. Spherical arm geometry positions the wrist through two rotations and one linear actuation. As in the previous cases, the orientation of the tool plate is achieved through three rotations in the wrist (A, roll; B, pitch; and C, yaw). In theory, the rotation about axis 2 could be 180 degrees or greater, and the waist rotation about axis 1 could be 360 degrees. Then if axis 3, robot reach, went from the retracted to the fully extended position, the volume of operating space defined would be two concentric half spheres. The actual work envelope for one of the most frequently used early spherical geometry machines is illustrated in Figure 2–13. Note that the working volume is much less than the theoretical volume of the machine in Figure 2–12. Again, this results from mechanical design constraints. Figure 2–14 is a small spherical geometry robot designed to be mounted on a numerical control machine tool to load and unload parts. Spherical geometry machines use either hydraulic or electric drives as the prime movers on the six axes, with pneumatic actuation used to open and close the gripper.

The advantages and disadvantages listed for cylindrical geometry can also be applied to spherical geometry, with the following exception: cylindrical geometry is more vertical in structure; spherical coordinates yield a low and long machine size. This is especially true for spherical machines that are designed to provide long horizontal reach.

Articulated Geometry

Articulated industrial robots, often called *jointed arm, revolute,* or *anthropomorphic* machines, have an irregular work envelope. This type of robot has two main variants; *vertically* articulated and *horizontally* articulated.

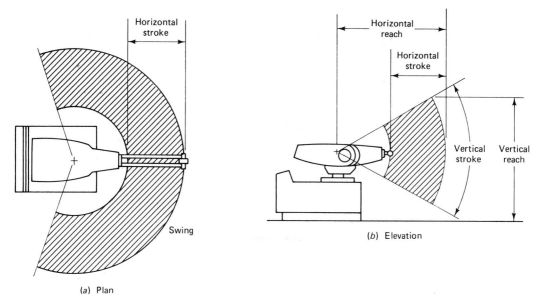

Figure 2–13 Westinghouse/Unimation 2000 Work Envelope.

The *vertically* articulated robot, sometimes called a jointed-spherical arm, Figure 2–15, has three major angular movements consisting of a base rotation (axis 1), shoulder (axis 2), and forearm (axis 3) joint. The irregular work envelope is illustrated in Figure 2–16. As in the previous arm designs, the orientation of the tool plate is provided by the three rotations in the wrist. Straight-line motion along any of the three coordinates, X, Y, or Z, requires the coordinated movement of a minimum of three joints; therefore, sophisticated controllers are generally required for this type of arm geometry. Electric drives with feedback control systems are used on most machines. The human-like movements of the jointed-spherical arm create the following advantages for robotic applications:

- Although it occupies a minimum of floor space, the robot achieves deep horizontal reach.
- A good size-to-reach ratio is achieved, a result of the arm's ability to fold up when in the retracted position.
- High positioning mobility of the end-of-arm tooling allows the arm to reach into enclosures and around obstructions.

This type of robot has one drawback: its more sophisticated control requirements result in higher cost for the machine. An example of a jointed-spherical configuration is illustrated in Figure 2–17. Note the *three-roll wrist* that generates the motion for all three wrist axes in a small mechanical package and *ball-screw* drive (covered by the flexible rubber sock) that lifts the arm.

There are two possible variations to the vertically articulated geometry just described: (1) an additional rotary motion axis (axis 4) in the forearm that rotates

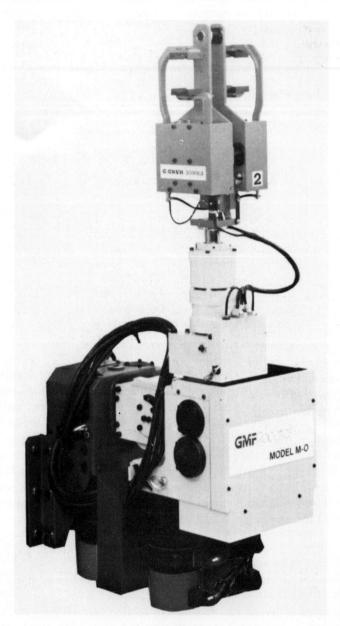

Figure 2–14 GMF Machine-tending Robot. (Courtesy of GMF Robotics)

the forearm link; and (2) an additional linear movement axis (axis 4) in the forearm that extends and retracts the forearm link.

The *horizontally* articulated robot has two position angular movements consisting of an arm and forearm rotation, and one position linear movement that is a vertical motion. Horizontally articulated arms are implemented with two mechanical configurations: (1) the *Selective Compliance Articulated Robot Arm (SCARA)* illustrated in Figure 2–18, and (2) the *horizontally base-jointed* arm.

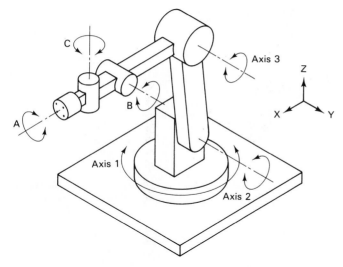

Figure 2–15 Jointed-spherical Geometry Robot.

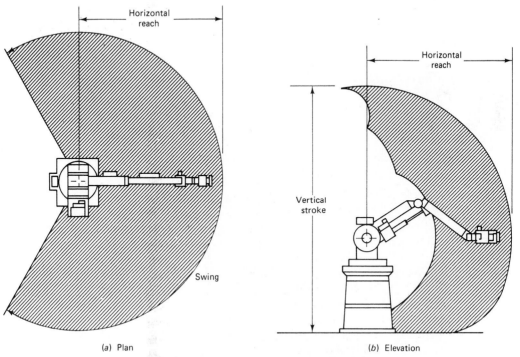

(a) Plan   (b) Elevation

Figure 2–16 Jointed-spherical Work Envelope.

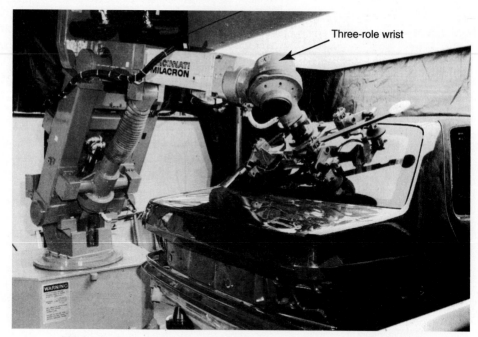

Figure 2–17 Jointed-spherical Robot Produced by Cincinnati Milacron Prior to Buyout by ABB.

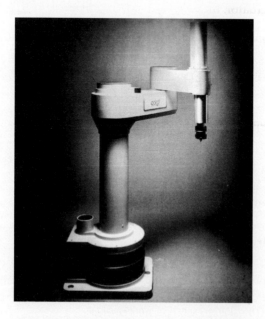

Figure 2–18 AdeptOne SCARA Robot. (Courtesy of Adept Technology, Inc.)

The SCARA machine (Figure 2–18) has two horizontally jointed arm segments fixed to a rigid vertical member. Positions within the cylindrical work envelope are achieved through changes in axes 1 and 2. Vertical movement of the gripper plate results from the Z axis located at the end of the arm. SCARA machines usually have only one wrist axis, rotation. This arm geometry is frequently used in electronic circuit board assembly applications because this geometry is particularly good at vertical part insertion.

The horizontally base-jointed arm uses the same construction as the SCARA with one exception: The vertical Z axis is located between the rigid vertical member and the shoulder joint of the upper part of the arm. With this configuration, the wrist and gripper mounting plate are located at the end of the forearm. Robots produced by the Reis Company made this cylindrical geometry popular.

2-3 POWER SOURCES

The three primary power sources used to drive manufacturing systems, namely, *hydraulics, pneumatics,* and *electromotive force,* are also used as prime movers in current robots. This section explains the classification of robotic arms by power source.

Hydraulic Drive

A basic robot hydraulic power system is illustrated in Figure 2–19; study the diagram carefully. The pump and tank provide oil at high pressure for the system; four-way control valves switch the flow of high-pressure oil, and one or more actuators produce the desired motion. The two actuators commonly used are the linear type for straight-line motion and the rotary type where rotational torque is desired. In the case of the linear actuator, the high-pressure hydraulic oil is forced into one end of the hydraulic cylinder. When the chamber fills, the high-pressure oil causes the piston to move; this movement forces the oil on the other side of the piston to flow out of the cylinder. This oil is returned to the pump through the four-way valve and return lines. In Figure 2–19 the solid arrows indicate the direction of oil flow and the resulting actuator movement. To reverse the actuator action, the fluid must flow in the direction of the dashed-line arrows. As indicated, the four-way valves, which control actuator movement, are driven by electrical signals from the robot control system. Figure 1–11 shows an early model hydraulic robot produced by Cincinnati Milacron (now ABB) with its five rotary actuators (labeled A through E) and one linear actuator (labeled F).

Hydraulic actuators have one primary advantage: a very high power-to-size ratio that affords large load capability. That single advantage is offset by several of the following disadvantages:

- Even the best hydraulic system will leak eventually.
- Hydraulic oil can become a fire hazard in arc welding applications.
- The additional equipment in the form of motor, pump, tank, and controls increases maintenance, energy, and robot costs.

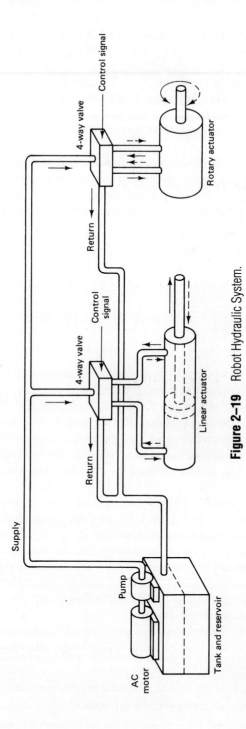

Figure 2–19 Robot Hydraulic System.

- A higher noise level is associated with hydraulic systems.
- Hydraulic maintenance skills are also required for repairs.
- Regular testing of the hydraulic fluid is required to determine the wear on actuators.

The increased payload capability of new electric robots has reduced the number of hydraulic robots used in manufacturing.

Pneumatic Drive

The basic system components in pneumatics are the same as those identified for the hydraulic system. The primary difference is that the power is being transferred by a gas under pressure rather than by oil. Study the section views of the linear pneumatic actuators and four-way control valves illustrated in Figure 2–20. The solenoid is an electrically actuated linear actuator that moves the round spool inside the valve against the spring. The two positions that the spool can assume permit air to be forced to either side of the piston in the robot actuator. The control values for the hydraulic robot described in the previous section operate in a similar fashion. In most robotic applications, the pneumatic actuators can be classified as linear or rotary. In each case the actuator has two usable positions, one at each end of the travel. The positions for linear devices are retracted and extended, and rotary actuators come to rest at the full clockwise rotation and the full counterclockwise position. Feedback control systems are rarely used to stop a pneumatic actuator between hard stops, because repeatability is not good. The repeatability is very good, however, if the actuator is driven against fixed stops at each extreme of travel. In some systems the fixed stops are adjustable slide blocks that stop the actuator as it extends or retracts. The gas serves to force the movable part of the actuator against the block; robot tool position is maintained by the block setting. A robot using this fixed-stop approach is called a *pick-and-place* device or

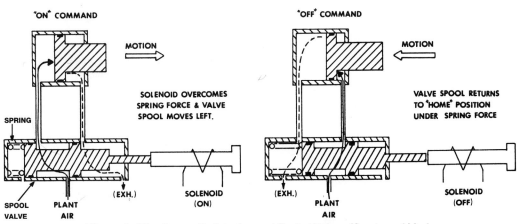

Figure 2–20 Pneumatic Actuators and Control Valves. (Courtesy of Mack Corporation)

bang-bang machine. The limited positioning capability of the pneumatic robot does not limit the use of this type of robot. In fact, more pneumatic robots are used in manufacturing worldwide than any other type.

The pneumatic power source has the following advantages:

- Compressed air is available in most manufacturing areas.
- Pneumatics is an inexpensive, well-developed technology.
- System leaks do not contaminate the work area.
- Fast operation and short cycle times are characteristic.

The primary disadvantage has been the inability to drive the pneumatic system using feedback control to provide proportional operation and multiple stops. Figures 2–4 and 2–11 show pneumatic-powered robots for small applications.

Electric Drive

The electric system includes a source of electric power and an electric motor. In most applications the motors are servomotors, but stepper motors are used on some robots where the payload is small. The servomotor can be dc or ac, with the latter becoming the more popular type. The electric drive is divided into two classifications: direct-drive and reduction-drive systems.

The electric motor with a *reduction-drive* system provides excellent rotational torque and angular positioning using a standard *belt* and *pulley, gear train,* or a *harmonic-drive* system. The motor and gears (A and B) in Figure 2–21 illustrate how the gear-type system would function; a belt-drive system is illustrated in Figure 2–22. These techniques are frequently used on jointed-spherical arm geometries because

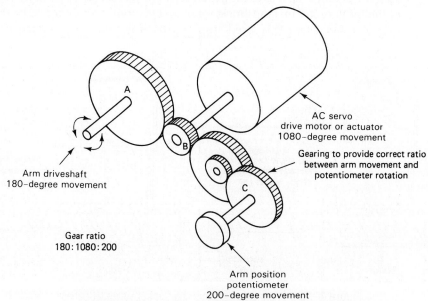

Figure 2–21 Position Feedback Potentiometer Gearing.

all the axes require angular-drive motors. When linear motion is required from an electric motor, a *ball-screw*-drive mechanism is used. Ball-screw drives (Figure 2–7) provide good mechanical advantage and positioning accuracy when linear motion must be produced from an angular drive source; however, ball-screws have a high noise level when operated at high speed.

The direct-drive system, such as that used in the AdeptOne robot in Figure 2–18, has the motor driveshaft connected directly to the axis. Degrees of rotation of the motor shaft and degrees of angular change in the axis have a one-to-one relationship. Consequently, direct-drive systems can produce only angular motion.

The advantages of the all-electric drive include the following:

- No generation of hydraulic or pneumatic power is required.
- No contamination of the work space occurs.
- Low noise level is maintained while operating.
- In addition, the direct-drive systems provide the fastest robot arm motion and quickest response.

Figure 2–22 Belt- and Pulley-drive System on End of Forearm of Jointed-spherical Robot to Produce Orientation Motion in Wrist.

The disadvantage is the limited lifting or payload capability of the electric system compared with its hydraulic counterpart. As a result, electrically powered robots are primarily designed for machine tending, material handling, part assembly, or welding and coating applications that have payload and end-of-arm tooling typically less than 250 pounds.

2-4 APPLICATION AREAS

It has been said that robotics is a solution looking for a problem. In the early days of work-cell design, robots were adapted as much as possible to solve the tasks at hand. The robot industry has now come of age, and robot manufacturers are zeroing in on target applications with machines whose characteristics match the specific job. The AdeptOne robot in Figure 2–18, for example, is a direct-drive robot with the speed and repeatability required to make robotic assembly cost-effective. Note the velocity and repeatability listed in Figure 1–17 for the AdeptOne robot. Applications for robots can be divided into two groups: *assembly* and *nonassembly*.

Assembly

Assembly remains one of the applications most difficult to adapt to robot automation. The key for robot assembly is a product design process that includes design for automated assembly as a primary design requirement. This can be a double-edged sword for robotics, however. A product designed for easy assembly is also much easier for humans to assemble. IBM, for example, redesigned its computer line for automated assembly. As a result, IBM found that it could improve profitability by taking many of the robots off the assembly line and replacing them with humans. Robot assembly is most effective when three assembly requirements exist: *high repeatability, small lot-size production,* and *varying part-size requirements.* If the assembly requires close tolerance but high-volume standard parts, then special-purpose high-speed assembly machines, called *fixed automation,* are the

Figure 2–23 Fixed Automation Assembly System for Radio Components Produced by METO-FER Corp. (Courtesy of METO-FER Corp., Pittsburgh, PA)

machines of choice. A carousel-type fixed automation machine with seven pneumatic robot positioners is illustrated in Figure 2–23. If the lot size is small or the parts to be assembled vary in size, then robotics is the answer.

Nonassembly

The primary applications in the nonassembly area have been *welding, spraying, coating, material handling,* and *machine loading* and *unloading.* These continue to be the major areas in the 1990s. The line drawn between assembly and nonassembly robots is certainly not a clear one, however, because assembly robots, like the AdeptOne, have been used in nonassembly tasks. Nevertheless, robot classification by applications, assembly or nonassembly, is valid because there are tasks at which each machine excels.

2-5 CONTROL TECHNIQUES

The type of control used to position the tooling separates robots into the categories of *servo* (closed-loop) or *nonservo* (open-loop) systems. An understanding of these two control techniques is fundamental to understanding the operation and programming of robots.

Closed-Loop Systems

A closed-loop control system measures and controls the position and velocity of the robot tool center point (TCP) at every point in the robot's motion throughout the work envelope. The closed-loop, or servo, system illustrated in Figure 2–24,

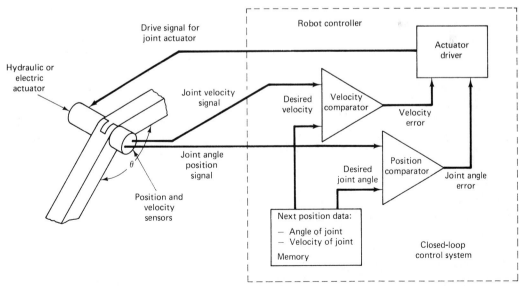

Figure 2–24 Closed-loop System.

uses information provided by sensors attached to the robot's movable axes to calculate the current position and velocity of the TCP. Study Figure 2–24 until you find the feedback sensor's signal path. Closed-loop controllers are much more complex and have a statistically higher chance to malfunction than open-loop systems; however, the advantages gained in the application of servo-controlled robots to certain manufacturing processes outweigh the problems created by the additional complexity inherent in the system. The advantages include these:

- Servo-controlled robots provide highly repeatable positioning anywhere inside the work envelope. This flexible, multiple-positioning characteristic is necessary for the implementation of robots in many production jobs.
- The servo type of controller with computer capability can provide system control for devices and machines that are external to the robot system. Many work cells require the synchronization of numerous peripheral devices, such as CNC machines, conveyors, gages, sensors, and readouts with robot motion. The servo type of controller usually has the input/output capability and the programming power to control the robot arm and the other support equipment in the work cell.
- Powerful programming commands are available to perform complex manufacturing tasks with very simple program steps. For example, palletizing any size part on any size pallet takes only a few commands.
- Interfacing robots to other computer-controlled systems, such as vision systems and host computers, is easier to accomplish with a servo-type controller.

Internal positioning devices mounted on the robot arm mechanically measure the angle of the joints or the linear movement of joints and convert these measurements into proportional electrical signals. In addition, other feedback devices on the arm provide rate-of-change data to be used in controlling the velocity and acceleration/deceleration rates at the TCP.

The three primary positional feedback devices are the *potentiometer*, the *resolver*, and the *optical encoder*. The potentiometer, or pot, is a variable resistor with a linear resistance and a movable wiper. The resolver uses magnetic coupling between transformers to measure rotation, and the encoders use an interrupted beam of light to determine position. In every case, the device is attached to the axis drive motor either directly or through reduction gearing. The rotational displacement of the drive motor is measured by the positional feedback device. A positional and rate-of-change resolver attached to the elbow joint of a robot is pictured in Figure 2–22.

Potentiometers.   The potentiometer is usually a single-turn type with gearing for reducing the multiple turns of the motor to a single turn of the potentiometer. The gearing is provided either by the robot manufacturer or is placed inside the potentiometer by its manufacturer, similar to standard ten-turn potentiometers. Figure 2–21 illustrates a simplified system with an ac servo, arm driveshaft, potentiometer gearing, and potentiometer. Assume that the arm drive rotates through 180 degrees for three turns of the motor, and the potentiometer has a rotation of 200

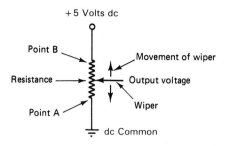

+5 Volts dc

Point B

Movement of wiper

Resistance

Output voltage

Point A

Wiper

dc Common

Potentiometer

Figure 2–25 Potentiometer.

degrees; then, a relationship exists between the position of the arm and that of the wiper on the potentiometer. Zero degrees on the arm would equate to zero degrees on the pot, so that the wiper would be in position A in Figure 2–25. The pot output would be zero volts because the wiper is connected to the ground. If the robot arm moves to the 180-degree point, then the pot wiper is rotated through 200 degrees of its 230-degree maximum rotation and would be in position B. The output voltage would be 4.3 volts. For arm positions between zero and 180 degrees, the potentiometer output voltage would be some value between zero and 4.3 volts. Without any error in the gears or the potentiometer, each degree of arm movement would cause the following change in output voltage:

$$\text{change in voltage output per degree} = \frac{\text{change in output voltage in volts}}{\text{change in shaft rotation in degrees}}$$

Even with the best resolution and accuracy possible in commercial potentiometers, the error introduced by the pot makes it difficult to use as the primary position feedback device. Instead, potentiometers are sometimes used for coarse position data in systems that combine potentiometers and optical encoders. In addition to lack of accuracy, the potentiometer has the following shortcomings:

■ Movement of the wiper contact on the resistance causes wear, which will eventually result in system failure.

■ Potentiometer output is affected by the environment.

■ The analog output of the potentiometer requires analog-to-digital conversion electronics to attain the necessary binary code (digital) for the controller computer.

Optical Encoders. *Absolute* and *incremental* types of optical encoders are frequently used as primary feedback devices to measure robot joint movement. In both types, a light source and light-sensitive receiver are separated by an encoder wheel, as illustrated in Figure 2–26. The beam of light from the source must pass through the encoder wheel to reach the light-sensitive receiver. The encoder in Figure 2–26 uses an opaque wheel with transparent regions, which pass the beam of light, and opaque sections, which interrupt the beam as the wheel is rotated. The disk can also be glass with opaque regions produced by depositing metal in

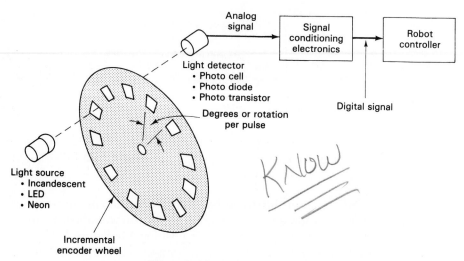

Figure 2–26 Incremental Encoder.

a precise pattern. The light source is either a light-emitting diode (LED) or a neon or tungsten lamp, which produce greater signal strength as a result of greater light output. The receiver is usually a photodiode or phototransistor, but photocells have been used in some encoders. The design illustrated in Figure 2–26 is called an *incremental-type* encoder because the receiver signal is a series of pulses produced as the light beam is interrupted. Note that the LED source is separated from the receiver by the disk. The pulses are produced when the beam, which is aligned with the transparent pattern, alternately passes through the wheel and then is blocked as rotation occurs. Each pulse represents the number of degrees or the increment of rotation necessary to produce it. The actual rotational position or starting position of the encoder shaft is not known; only the number of increments changed from the previous position is provided.

The typical pulse output waveforms from this type of encoder are illustrated in Figure 2–27. The waveform on the top in Figure 2–27 is the actual output from the receiver on the encoder, and the waveform on the bottom is the pulse train generated by the encoder electronics. Each time that the raw pulse crosses a predetermined value, a single square pulse is produced electronically.

In addition, the incremental encoder in Figure 2–26 does not provide direction information, since the same pulse train, Figure 2–27, will be produced regardless of the direction of rotation. Direction information is included when a second light beam is added to the system. Figure 2–28 shows an incremental encoder with direction sensing added. The two rows of marks are slightly offset so that for counterclockwise rotation the first beam is interrupted before the second beam is broken by the inner track. When the disk is rotating clockwise, beam two is broken first, and then beam one. A digital circuit determines which beam is leading the other so that direction information is available. Using the time Tp (pulse period) in Figure 2–27, the controller electronics determine the rate of change so that joint

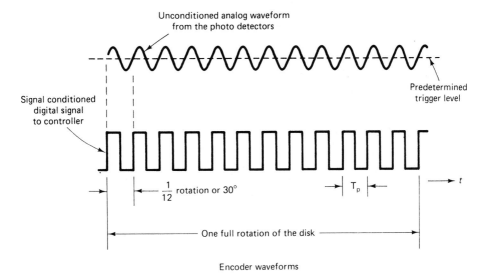

Encoder waveforms

Figure 2–27 Encoder Waveforms.

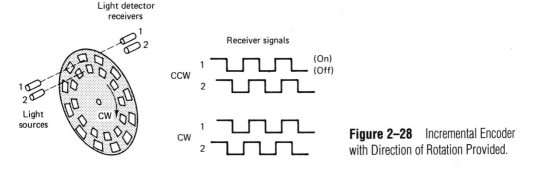

Figure 2–28 Incremental Encoder with Direction of Rotation Provided.

velocity and acceleration/deceleration data are known. Note also that the resolution or accuracy of the encoder is directly proportional to the number of marks (or holes) on (or in) the disk, which corresponds to the number of pulses per revolution. For a resolution of 0.1 degrees the disk would need 3600 marks.

An *absolute-type* of optical encoder eliminates the primary shortcoming of the incremental system by providing actual rotational position information. This improvement is accomplished by producing a multidigit binary code for every incremental change in the disk position. Figure 2–29 shows a 4-bit absolute encoder. Note that each bit requires its own ring of transparent patterns and a separate light source and receiver for every bit of code or track on the disk. The resolution on the 4-bit encoder is poor, as only 16 codes or different positions of the encoder disk are possible. The resolution in degrees for 16 positions is as follows:

$$\frac{360 \text{ degrees}}{16 \text{ different codes}} = \frac{22.5 \text{ degrees}}{\text{code}}$$

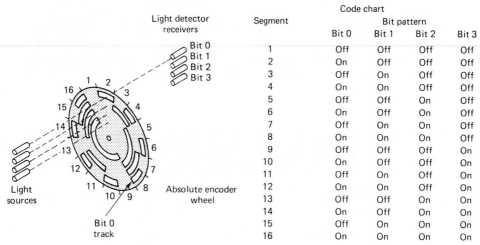

Segment	Code chart Bit pattern			
	Bit 0	Bit 1	Bit 2	Bit 3
1	Off	Off	Off	Off
2	On	Off	Off	Off
3	Off	On	Off	Off
4	On	On	Off	Off
5	Off	Off	On	Off
6	On	Off	On	Off
7	Off	On	On	Off
8	On	On	On	Off
9	Off	Off	Off	On
10	On	Off	Off	On
11	Off	On	Off	On
12	On	On	Off	On
13	Off	Off	On	On
14	On	Off	On	On
15	Off	On	On	On
16	On	On	On	On

Figure 2–29   Absolute Encoder with Code.

Detection of a small rotational change will require a disk with many more tracks and a large number of bits in the code. The code produced by the disk in Figure 2–29 is outlined in the adjacent table. The rotational position of the shaft indicates the code on/off pattern, and the direction is determined by the increasing or decreasing order of the patterns generated.

Both types of optical encoders are used in robot control systems because the encoders have a number of advantages over the potentiometer described earlier:

- Greater resolution and accuracy
- Noncontact measurements, which reduce wear and improve reliability
- No loss of resolution in the conversion of encoder output to controller-compatible code (eg, a 10-bit absolute encoder will have a resolution of 10 bits).

The two disadvantages encoders have when compared with the potentiometer type of positional measuring sensors are higher cost and larger physical size, especially absolute encoders with high resolution.

Resolvers and Synchros. A synchro is another type of device used for internal control of servo-driven robot arms. The word *synchro* is a generic term covering a range of ac electromechanical devices that are basically variable transformers. Figure 2–30 shows the schematic diagram and wiring of the two basic types frequently used. The synchro has three windings mounted at 120-degree intervals around a stationary stator; the resolver has two windings offset by 90 degrees around its stationary stator. Both types have either a single or a double winding, called the rotor, which rotates inside the stator windings. The degree of magnetic coupling between the rotor and stator windings varies with the shaft angle or amount of rotation present in the rotor. If an ac reference voltage is applied to the rotor winding, the stator windings have an ac output that is a function of the cou-

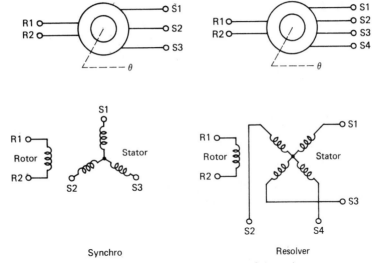

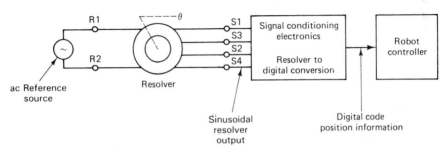

Figure 2–30 Synchro and Resolver Schematics.

Figure 2–31 Block Diagram of Resolver-based Position System.

pling coefficient or shaft angle of the rotor. The amplitude of the ac output present on the stator windings will vary as a function of the shaft angle and will be represented by a sine wave on one output and a cosine wave on the other. To interface the output of the resolver with the digital computer electronics in a robot controller, the analog position data must be converted into binary code. The three widely used methods of conversion are (1) time-phase shift, (2) sample and hold, and (3) tracking. The basic block diagram for a resolver-based position encoder is given in Figure 2–31. Resolver position sensors are identified in the robot pictures in Figure 2–7 and 2–22.

A comparison of the resolver-based position encoder to the optical encoder indicates the following advantages for the resolver-based system:

■ The optical encoder is more dependent on mechanical precision to achieve positional accuracy, whereas the resolver relies more on electronic circuits for the same accuracy.

- In the resolver, the conversion electronics are remote from the resolver unit itself. Thus, the electronics unit can receive greater protection from environmental conditions. In the optical encoder, the conversion electronics are inside the unit because signal strength from the optical unit is low.
- The resolver-based system is less susceptible to damage because no glass disks or light sources can burn out.
- The resolver has smaller physical size at higher resolution values.
- The resolver requires fewer output wires between the sensor and controller for absolute encoding.
- The position sensing of the resolver is always absolute.

When compared with the optical encoders, the resolver-based system has the following disadvantages:

- It is more expensive than an incremental encoded system, especially at higher resolutions.
- It needs an ac reference voltage.
- The conversion electronics are more expensive because the conversion unit must have 1 to 2 bits of resolution greater than that needed by the system as a whole.

Most servo-controlled robot systems use either optical or resolver-based positional encoders to determine the present joint angles for use in controller calculations. Some robots, however, use a combination of potentiometers and optical encoders for position information. Potentiometers are used for coarse location of the joints, and incremental encoders are used for fine-positioning data. In general, optical encoders have greater application at lower resolution values, and resolvers are used when greater accuracy and higher resolution are necessary.

Operation. The closed-loop robot system in Figure 2–24 could be implemented with any of the position sensing devices just described. In operation, the position of the robot arm is continuously monitored by a position sensor, and the power to the actuator is continuously altered so that the movement of the arm conforms to the desired path in both direction and velocity.

A six-axes robot arm would have six joints and closed-loop circuits like that shown in Figure 2–24. The controller gets joint-angle information from a position sensor that could be a potentiometer, encoder, or resolver. Each of these position sensors can provide a continuous electrical signal that represents the current angle of the joint. Thus, the controller can compare the current position to the next desired joint angle, which is calculated from the data stored in the controller memory when the robot was programmed. This electronic comparator will produce a joint-angle error signal that can be used by the actuator driver to change the joint angle so it will agree with the desired value.

In a similar manner, the rate of change of the joint angle is continuously monitored by the velocity sensor, which is usually a tachometer measuring the revolutions per minute (rpm) of the rotary actuator. The rate at which the joint is moving is directly proportional to the rpm of the rotary actuator driving the joint. The rate of change of the joint is compared with the desired velocity by a separate comparator and produces a rate of change or velocity error. The error from the two comparators is analyzed by the actuator driver electronics to determine the velocity and direction the actuator should assume for the joint to reach the value required for the next position in the program.

If a machine has 6 degrees of freedom or six axes, then the position of the joint in each degree of freedom is individually measured, and the electrical signals are sent by wire to the controller for analysis by the servo electronics. This sampling of joint positions and corresponding corrective action to the actuator occurs many times a second so that the visual effect is a smooth movement in the robot arm from one position to another.

The servo, or closed-loop, system is used in any application in which path control is required, such as in welding, coating, and assembly operations. Operation of servo-driven robots is often more difficult than that of their nonservo counterparts. System start-up involves a series of keystrokes from the keyboard and/or teach pendant, followed by the teaching process that satisfies the application. After testing and debugging the application program in the controller memory, the program is usually saved on disk or magnetic tape as a backup. The program will remain in the robot controller memory as long as the main power is not interrupted and the controller backup battery remains charged. The system is ready to execute the current program stored in memory whenever the robot is powered up.

Open-Loop Systems

A *nonservo*, or *open-loop*, system (illustrated in Figure 2–32) provides no information to the controller concerning arm position or rate of change. As the system drawing indicates, the arm receives only the drive signal, and the controller relies on the reliable operation of the actuators to move the arm to the programmed location. Open-loop systems account for more than half of current robot applications in the United States and nearly 65 percent of those abroad. There are several advantages of an open-loop system:

- Lower initial investment in robot hardware
- Well-established controller technology in the form of programmable logic controllers
- Less sophisticated mechanical and electronic systems with fewer maintenance and service requirements
- A larger pool of engineers and technicians familiar with the requirements of this type of system

The control of nonservo robots includes internal mechanisms for establishing accurate positioning and external drive devices for ensuring the required sequence of movements.

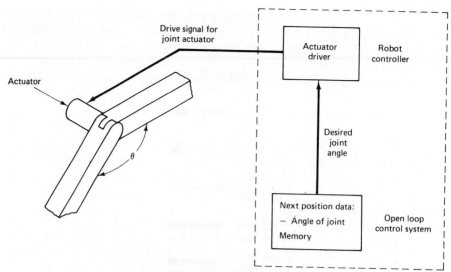

Figure 2–32 Open-loop System.

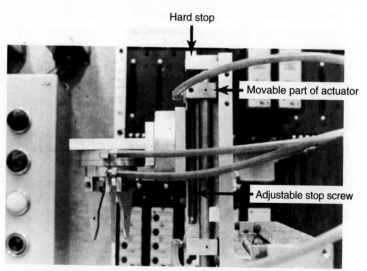

Figure 2–33 Adjustable Hard Stops on B-A-S-E robot. (Courtesy of Mack Corporation)

Internal Positioning Devices. The elimination of continuous position information on arm position requires that *internal positioning mechanisms* be used to accurately position the robot arm. The most frequently used position mechanisms for nonservo robots include the following four:

1. *Fixed hard stops* limit the movement of pneumatic or hydraulic actuators at each end of their travel. These stops can either coincide with the natural

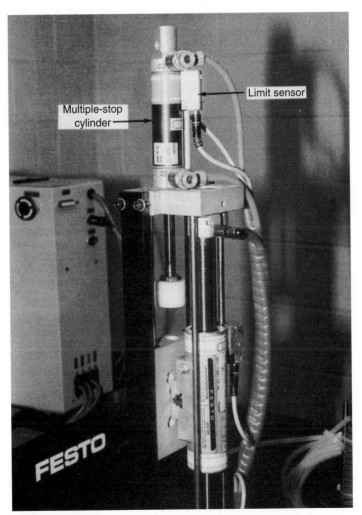

Figure 2–34 Limit Sensing and Multiple Stops on Festo Root.

extension length of the actuator or can be added to the actuator to limit travel.

2. *Adjustable hard stops* limit the movement of pneumatic or hydraulic actuators at each end of their travel. These stops are identical to the fixed hard stops except that they are adjustable over part of the length of the actuator extension. Figure 2–33 shows a threaded shaft on a Mack modular robot used to adjust the stop on the Y transporter. The threaded adjustment screw positions the movable part of the Y transporter away from the hard end stop. Figure 2–34 illustrates how two different Z axis stops are produced on a Festo pneumatic robot using an additional pneumatic cylinder. Note that the stops are selected under program control using the smaller pneumatic cylinder on the top of the Z axis travel. When the small cylinder

at the top of the Z axis is extended, the upward movement of the arm is stopped at a lower elevation.

3. *Limit switches* produce variable stops on nonservo actuators. The robot pictured in Figure 2–35 shows the limit switches used to produce variable degrees of rotation of the arm on the base.

4. *Stepper motors* produce a fixed degree of rotation based on the number of pulses that are applied to the motor windings. Stepper motors are not often used on industrial robots, but they are found in peripheral hardware, such as positioning tables.

The rugged nature of the internal position control devices assures long operating life and high mean-time-between-failure (MTBF). The pneumatic systems require only clean dry air, line filters, and an oiler to provide trouble-free performance. Some pneumatic systems are prelubricated for life so that oilers are not necessary.

Of course, hydraulic systems require additional maintenance as a result of the pump and tank, which are part of all high-pressure oil systems. With regular filter changes and chemical tests of the hydraulic fluid, however, the nonservo hydraulic robots provide a good MTBF.

Operation.   The operation of nonservo robot systems is similar to that of the servo type except that the number of programmed points is limited to the fixed or variable stops present on each actuator. In addition, the controller does not know the position of the arm and tool while the robot is moving from one point to another. On every axis, however, there is a *fixed stop,* or *limit,* at each extreme of travel that provides the positioning accuracy at that point. The joint will only stop moving when it reaches one of the two possible extremes of travel or when the actuator driver removes the drive signal after a predetermined amount of time. A

Figure 2–35   Multiple Stops on Prab Nonservo Feedback System. (Courtesy of Prab Robots, Inc.)

signal from the controller causes the power source to drive each axis actuator from the present limit or stop to the opposite limit.

In a technique known as *limit sensing*, a sensor is placed at the limit of arm travel to verify that the new position or arm limit has been reached. For example, if a nonservo robot is used to unload a die-casting machine, the system design would require verification that the robot gripper and the part held in the gripper are both clear of the dies before the die-casting machine would be cycled and the dies would be closed. Therefore, limit sensing would be required on the axes used to move away from the die-cast machine. After the controller drives the arm away from the machine, a verification from the limit switch signals the controller to continue in the program. Limit sensors are visible on the Festo robot in Figure 2–34. Although limit sensing is used in some applications, many robots use the integrity of the mechanical system and the time allowed to reach the new position to assure accurate operation.

The open-loop system is referred to in the industry as a *stop-to-stop* or *pick-and-place robot*. A pneumatically powered modular component system is pictured in Figure 2–4. The system is assembled by the user from a wide selection of linear and rotary actuators. The advantages of the nonservo robot are the simplicity of the system and the resulting reliability. Pneumatic robots or pneumatic-actuated devices are successfully used in every manufacturing area. The disadvantages for the nonservo stop-to-stop type are its limited number of fixed-stop locations and its inability to handle complex manufacturing tasks. Despite any disadvantages, the low initial cost makes this system an attractive choice for machine load-and-unload applications.

2-6 PATH CONTROL

The least ambiguous method for classifying robots is based on the type of path control that the controller provides. Path control is a way of defining the method that the robot controller is using to guide the tooling through the many points in the desired arm trajectory. The four types of path control from least complex to most complex are *stop-to-stop, continuous, point-to-point,* and *controlled path*. Some segments of the robotics industry identify stop-to-stop machines as point-to-point without servo feedback; however, except for this one deviation the industry is consistent.

Before describing the four types of path control, a clear understanding of how the controller is storing the program points is important. The programmed points in a robot program are called translations or position points. Figure 2–36 shows a simplified robot with 4 degrees of freedom programmed to grasp a part. The robot is shown in the three positions through which it must pass to complete the pickup of the part. Point 1 is the starting point; point 2 positions the gripper over the part; and point 3 puts the gripper in a position where the parallel jaws can close to secure the part. For this operation the three *translation* points represent the minimum number of program steps that would have to be used. The robot operating system would record in the controller memory the necessary information about the path to permit the robot to return to these three points as the program is executed. The

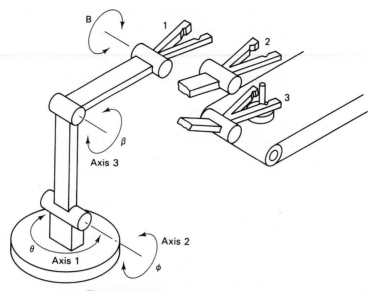

Figure 2–36 Point-to-point Path Control.

minimum information necessary to permit the robot to return to any point in its work envelope is the position of each degree of freedom or of each axis when the robot is at the point. In the case of joints that move through rotation, the angle is the critical value, and in the case of linear movement the distance extended would have to be preserved. The actual data stored in memory for each programmed point vary, depending on the type of path control present and the vendor's design. The data take three forms: joint angles in angular units, Cartesian space coordinate values, or pneumatic valve states, such as *on* or *off*.

Most often continuous-path–type machines store the actual joint angles for each programmed point. Figure 2–37 shows a portion of the controller memory of a continuous-path system with joint angles for three recorded points. For point-to-point servo machines the system stores the Cartesian space coordinates of the programmed points. For example, the X, Y, Z, A, B, and C values would be saved. The coordinate values are used to calculate joint angles or actuator extensions during program execution. In actual practice the information or numerical values in memory would be recorded as binary (1s and 0s) numbers. Therefore, regardless of the path control in use by a robot system the problem is reduced to recording the critical information about every location in the work envelope that the robot must pass through in the process of executing a series of moves. This basic process applies to each type of path control discussed in this chapter.

Stop-to-Stop Control

In *stop-to-stop path control* the robot system is operating open-loop, which means that when the axis moves the position and velocity of the axis is not known to the controller. An example will demonstrate this. The robot in Figure 2–38 must

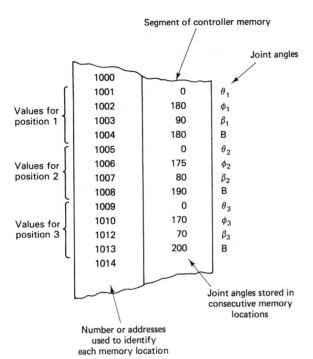

Segment of controller memory

Joint angles

1000		
1001	0	θ_1
1002	180	ϕ_1
1003	90	β_1
1004	180	B
1005	0	θ_2
1006	175	ϕ_2
1007	80	β_2
1008	190	B
1009	0	θ_3
1010	170	ϕ_3
1012	70	β_3
1013	200	B
1014		

Values for position 1

Values for position 2

Values for position 3

Joint angles stored in consecutive memory locations

Number or addresses used to identify each memory location

Figure 2–37   Segment of Robot Controller Memory.

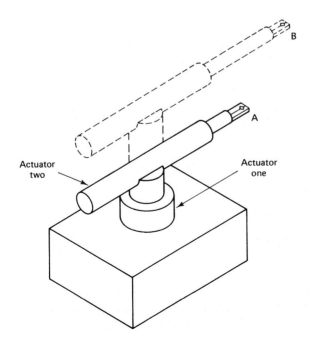

B

A

Actuator two

Actuator one

Figure 2–38   Stop-to-stop Path Control.

go from point A to point B as part of a programmed move. Because the motion of the actuators is not sampled with a feedback system, the actual position of the axis is not known until the actuator is driven to the desired limit or mechanical stops in the actuator cylinders. As a result, the only information stored in memory is a sequential list of *on/off* commands for each actuator driver. In this case the memory would have *on* for actuator one followed by *on* for actuator two.

Point-to-Point Control

The primary programming device in *point-to-point* controllers is the *handheld teach pendant*. The controls on the pendant for a Sankyo robot controller are shown in Figure 1 –16. Teach pendants usually have two switches for each degree of freedom on the robot. The programmer can move each axis independently in either direction with these controls. In addition, the pendant provides a programming button that commands the controller to record in memory the current position of every joint or axis on the robot. A button is also provided for emergency stop and single stepping through a program. These functions are common among the teach pendants of most manufacturers; often, additional control switches are provided for special functions of a machine, such as speed control during programming or gripper open and close.

The tooling or gripper is moved into position by the human programmer using the teach pendant. Then the robot controller is commanded to record in memory the coordinate values for the position of the arm and tooling. This information represents one point in a programmed move. The same procedure is followed at the next desired point in the work cell. For every programmed point the X, Y, Z, A, B, and C values are recorded by the system when the record button is depressed. This process is repeated until all the desired positions of the tool have been stored in the controller memory in the form of Cartesian space coordinates. It is important to note that the path that the robot takes as the programmer moves the tool from one desired position to another will have no effect on the final program path when the program is run. Figure 2–39 illustrates this concept with the positions that would be required for a point-to-point path controller to move a part from one conveyer to another. The following four points would be required: pick up the part at point 1; raise it to point 2; transfer it to point 3; and lower it to point 4. The solid lines represent the desired path of movement when the program is executed by the controller. The dotted line represents the path that the part might have taken when the programmer was teaching the robot the desired operation. *The translation points marked 1 to 4 are the only points that were recorded in the controller memory by the programmer,* so the deviation from the desired path as the programmer moved the tool from points 1 to 4 did not affect the final operation.

Point-to-point controllers can store thousands of program points for a complex application and can also store several different programs; thus a robot can adapt its movement to different manufactured parts or conditions in the work cell. For example, in an inspection operation the controller could drive the gripper through a programmed routine that moves parts through a laser gauge and onto an output conveyer. As soon as a part fails inspection, however, the process must

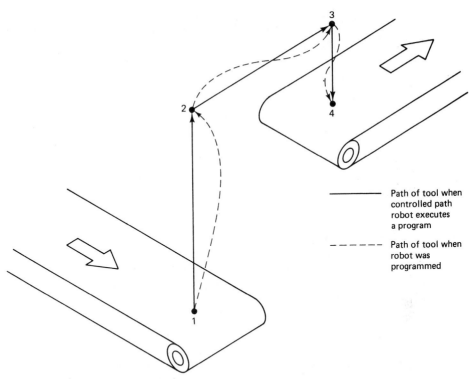

Figure 2–39 Point-to-point Programming Example.

Path of tool when controlled path robot executes a program

Path of tool when robot was programmed

be altered to drop the part into a rework bin. This is accomplished by activating a second stored program that has the set of points for the path to the rework bin. This process is called *branching*.

The point-to-point–type system provides feedback control of each axis over the entire path, with the controller driving each axis from its starting angle or position to the angle required for the next programmed point along the path. Figure 2–40 shows a symbolic spherical robot moving from point 1 to point 2, where the first and second points are the same distance from the robot, but point 2 is higher. The controller must change 4 degrees of freedom in this example. The base must rotate through angle θ; the shoulder must go from angle $\phi 1$ to angle $\phi 2$; the arm must extend along Z to reach the higher point; and the wrist must pitch from angle $\gamma 1$ to $\gamma 2$. The controller will change each axis at its maximum rate; therefore, the shoulder and pitch actuators, with the least change, will complete the move well in advance of the base actuator, which requires the greatest travel distance. This causes the end effector to assume the non-straight-line path indicated by the dotted line in Figure 2–40. This lack of path control presents no problems with robots used in applications such as material handling or machine tending where the straight-line path is not required; however, it could cause programming difficulty in arc welding situations in which the robot is required to follow a seam between two metal parts. To achieve straight-line motion with a point-to-point controller of

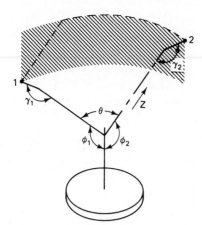

Figure 2–40 Example of Point-to-point Non–straight-line Motion.

this type many points must be programmed very close together to keep the changes in every axis small and about equal.

The advantage of point-to-point–type control is that relatively large and complex programs can be obtained with a system that is moderate in cost yet has proven reliability. Although there are many applications in which path control is not a requirement, the lack of straight-line control does limit the number of applications of a point-to-point system and would thus be considered its primary disadvantage. Some robot systems have regions in the work envelope where the programmed path is the point-to-point method just described or the path control technique covered in the next section.

There is a good selection of servo-controlled point-to-point robots available with either electric or hydraulic drives and with any desired arm geometry; however, the cylindrical type is the most popular. The lifting capacity varies from several kilograms to more than 900 kilograms with repeatabilities as precise as ± 0.05 millimeters.

Controlled Path

The *controlled path* robot is a point-to-point system with added capability to provide control of the end effector or TCP as it moves from one program point to another. The system is programmed in the same manner as the standard point-to-point machines with each point in the path recorded using the teach pendant. The difference occurs when the program is executed; the primary distinction is the *straight-line motion* between the programmed points. The axis actuators are driven in a proportional manner, with the axis requiring the most change driven faster than the axis requiring the least change. One result is that a straight path is followed between programmed points without any additional programmed points. In addition, the *velocity* between points can be specified in the program, along with special tool moves such as an arc welding weave pattern. All this capability is a result of increased controller intelligence and has an added benefit of enhanced programming features such as program editing, trouble diagnostics, larger memory capability, and added end-effector control.

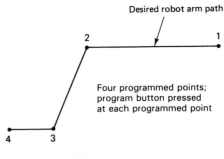

Desired robot arm path

2   1

Four programmed points;
program button pressed
at each programmed point

4   3

(a) Point–to–point type control

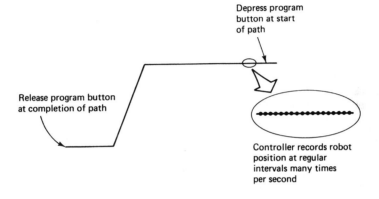

Depress program
button at start
of path

Release program button
at completion of path

Controller records robot
position at regular
intervals many times
per second

(b) Continuous–path control

Figure 2–41   Comparison of Point-to-point and Continuous-path–type Control.

Continuous Path

The primary difference between point-to-point controlled path just described and *continuous-path* control is the number of programmed points that are saved in controller memory and the method used to save them. Study the programmed path in Figure 2–41 to see the difference. The point-to-point type of robot would require just four programmed points stored in memory to record this arm motion. The continuous-path machine would store hundreds of points for the same arm motion, however. The reason for the large difference in the number of points stored is the method used by each type of system to record a programmed path. In point-to-point programming the programmer moves the robot to the desired location and presses the program button. A single point in the programmed path is saved. After a sequence of these, the program is complete. The point-to-point controller then drives the robot from one point to the next. By contrast, in continuous-path programming the programmer teaches the desired path using a teach stand or by physically moving the robot arm. The program button is depressed at the start of the move and not released until the desired path is completed. While the program button is depressed, the continuous-path controller is recording program points in memory at the rate of six or more per second.

The continuous-path programming process records every move that the programmer makes as the arm or teach stand is moved. The point-to-point process, however, records only the location of the arm when the program button is pressed. Any motion in the arm when it is moved from one point to the next is not recorded. Continuous-path controllers are very useful in applications such as paint spraying in which the robot's motion must duplicate the skill of the operator.

2-7 CLASSIFICATION BY THE INTERNATIONAL STANDARDS ORGANIZATION

The ISO has issued several standards documents to assist in the collection of valid robot data. The International Standard for Industrial Classification (ISIC) codes help target specific industries for data collection, while standard ISO/TR/8373 provides an extensive list of robotic terms and definitions. In addition, the standard classifies robots into four areas: *sequenced, trajectory, adaptive,* and *teleoperated*. The operation of the robot controller provides the primary difference in the classification categories named in the standard.

Sequenced

The nonservo-controlled pneumatic robot with stop-to-stop path control in either a Cartesian or cylindrical geometry best describes this category. The on/off, or binary, nature of the controller output drives the axes in a sequential fashion to well-defined end points. The trajectory or path, however, is not controlled or defined. The most frequently used controller for this category of robots is a programmable logic controller.

Trajectory

This category includes all robot geometries with servo-driven electric or hydraulic axes and controlled-path trajectory operation. The multi-axes move and internally generated straight-line motion characterize this classification.

Adaptive

This new category includes "thinking machines." Pure examples of machines in this category do not exist at present; however, trajectory-operated robot systems with sensory, adaptive, or learning control functions typify robot systems moving toward this classification.

Teleoperated

Teleoperated robots which extend the human sensory-motor functions to remote locations have been used for many years to handle radioactive material. This category includes a new class of teleoperated machines that can be programmed to respond to the operator's actions.

2-8 SUMMARY

The objective of this chapter was to introduce the general concept of robot classification and to provide an overview of all types of robot systems. Industrial robots can be grouped by several techniques that include arm geometry, power source for the joint actuators, intended applications, the presence or absence of feedback, the type of path control, and ISO standards.

The arm geometries currently available are Cartesian, cylindrical, spherical, and articulated. These mechanical configurations can use pneumatic, hydraulic, or electric drives as a source of power for the joints or axes that are moved. Application is not a good classification technique, but generally the areas of painting-coating, arc welding, assembly, and machine tending have received special attention from the robot industry. A much better classification scheme designates a robot system as either servo or nonservo controlled, which means that the system either knows the current position of every axis through the use of feedback signals or it does not sample the current position values and has no feedback. In the closed-loop, or servo, type of robot systems the joint angles and arm extensions are most often measured by potentiometers, optical encoders, or resolvers. The optical encoders used include both the incremental and absolute types. Most robot systems use either optical encoders or resolvers, because the accuracy and reliability of potentiometers is not adequate for industrial applications. Joint position sensors are part of the internal mechanisms that control the motion of the robot arm. In nonservo, or open-loop, systems, the limit of motion is set by mechanical stops on the arm. The internal limit devices are usually fixed hard stops, adjustable stops, or programmed adjustable stops. Another good classification technique uses path control to distinguish between current robots. The path control types include point-to-point, controlled path point-to-point, continuous, and stop-to-stop, with the last type using open-loop control in most situations.

QUESTIONS

1. Compare the five basic robot geometries according to the advantages and disadvantages of each arm, work envelope, typical applications, and power sources.
2. Compare the three types of drive power sources by preparing a table that illustrates the strengths and weaknesses of each.
3. Make a list of the most important machine characteristics that should be incorporated into an assembly robot system.
4. What is the primary difference between a servo and nonservo robot system?
5. What are the three primary sensors used in servo feedback loops, and how do they measure the joint positions?
6. What two parameters are measured at every joint of most servo-controlled robots?
7. What are the advantages associated with nonservo robots?

8. Describe four techniques used on nonservo robot arms to achieve positioning.

9. What are the maintenance requirements on hydraulic and pneumatic nonservo robots?

10. Describe four advantages of servo-controlled robots.

11. Describe four limitations present when potentiometers are used for feedback sensors.

12. Describe how incremental and absolute optical encoders are constructed and how they measure shaft position.

13. Describe how the direction of the rotation is determined in each type of optical encoder.

14. What are the advantages and disadvantages of optical encoders compared with potentiometers?

15. What is a synchro?

16. What is the difference between a synchro and a resolver?

17. How do resolvers operate when used as shaft angle encoders?

18. What are the advantages and disadvantages of resolver position detection compared with optical type encoders?

19. Describe the three program arm control techniques used in servo robot arms.

20. What types of manufacturing applications would be best served by a nonservo system?

21. What characteristics of manufacturing applications require that the robot used have a closed-loop system?

22. Why does the continuous-path system have a limitation on the length and number of programs that can be stored?

23. Why is the continuous-path system ideal for applications such as paint spraying and coating?

24. List three typical applications for each of the four types of path control. Try to match job tasks with the strengths of each type of path control.

25. How does the SCARA arm geometry differ from vertical articulated geometry?

PROBLEMS

1. Determine the volts per degree of rotation for the feedback potentiometer in Figure 2–25. Assume that a zero driveshaft position corresponds to zero output volts on the potentiometer and a total rotation of 230 degrees is possible. Determine the current position of the axis driveshaft and the drive motor in degrees when the potentiometer has an output voltage of 3.7 volts.

2. What is the rotational resolution in degrees of an absolute encoder with a 12-bit binary code? (2 to the 12th power different codes will be present)

3. How many optical marks will be required on an incremental encoder with a resolution of 0.05 degrees?

PROJECTS

1. Use primary and secondary research to make a list of robot applications at the companies identified in project 2 in Chapter 1. (If a single company has a large number of robot applications, just focus on that company.) Include the classification of each robot using the classification techniques (arm geometry, power source for the joint actuators, applications, servo/nonservo, brand) described in this chapter.

2. Use dBASE, Access, or another personal computer database software package to build a robot selection database. Include the following robot data: work envelope limits, payload, speed, positioning accuracy, repeatability, geometry, program control techniques, actuator power source, special features, and any other parameter required by your instructor. Set up queries that permit searches for robots based on single parameters or combinations of parameters. Use robot manufacturer's data sheets provided by your instructor and specifications obtained directly from robot vendors.

Automated Work Cells
and CIM Systems

3-1 INTRODUCTION

The ultimate goal of the enterprise is to develop an internal strategy that raises manufacturing performance to a level higher than that of the competition. The strategy often includes the development of production work cells with many forms of automation, including robots. In Chapter 1, order-winning criteria were used to identify and rank performance measures most critical for manufacturing success in a given market. To justify the investment in automation, the production cell must support some combination of the following performance measures: lower manufacturing cost, higher product quality, better production control, better customer responsiveness, reduced inventory, greater flexibility, and smaller lot-size production. In the 1970s the focus was on designing production work cells that were islands of automation. These stand-alone production cells were essentially isolated from other automated cells and the rest of the production system. Although the automated cells frequently improved the production performance, they fell short of the performance promised by the automation equipment vendors.

The disappointing performance was often traced to three sources. First, the automated cell was isolated from the rest of the production system. History has proved that complex, highly integrated production systems—called flexible manufacturing systems—are often difficult to maintain and operate; nevertheless, some integration of the production data and products of the automated cell with the rest of the production system is often beneficial. The level of integration is a function of the end product and the production hardware and software implemented.

The second cause for the low performance of the automated cell resulted from deficiencies in the work-cell design process. The early automated cells were usually manual manufacturing operations that were changed to automated production. In the conversion process, the manual functions were just replaced by the automated equipment. For example, when a robot replaced a human operator who loaded and unloaded parts from a machine, the duplication of the human motion was the major focus of the cell design process. As a result, the automated cell incorporated many of the poor production practices present in the manual cell. The third cause of low performance in early automated production cells was insufficient training and preparation of people to set up and maintain the systems.

A study of the failures in early automation efforts led to the development of the process called CIM to support effective integration of enterprise systems. Each company approaches the integration of the production systems, product data, and management of the organization differently. Even though every automation project is unique, agreement exists on the process that should be used by organizations to respond to the external challenges present in the marketplace. The CIM process is effective for projects as small as a single automated cell with a robot and production machine or as large as an automotive assembly plant. This chapter describes the general process for integration of automated systems and the specific implementation process for robotic work cells. A good starting point for our study is the three-step CIM implementation process.

3-2 THE CIM IMPLEMENTATION PROCESS

The implementation of a successful CIM system follows a three-step process that includes *assessment, simplification,* and *implementation.* In the first step of the CIM design process, the organization is studied to determine strengths and weaknesses. After the initial assessment, a simplification process is applied to eliminate all waste from the manufacturing area(s) to be automated. The final step in the CIM design process is the acquisition and implementation of the hardware and software for the production cells or manufacturing systems. A detailed description of the three-step process follows.

Step 1: Assessment of Enterprise Technology, Human Resources, and Systems

Building a work cell or integrated production system that uses automation technology such as robotics requires extensive planning, many months of hard work, and a substantial investment in people, hardware, and software. The first step in the system design process must be assessment, because a thorough understanding of the current operation is critical for success in the simplification and implementation steps. The assessment of the enterprise's current *technology, human resources,* and *systems* includes a study to determine these factors:

■ The current level of technology and process sophistication used in manufacturing

- The degree of employee readiness for the adoption of CIM automation across the enterprise
- The reason *why* the production systems function as they do

In each case the capability, strengths, and weaknesses are checked and documented.

The assessment process is an internal self-study with a large educational component. The critical nature of education is illustrated by the bar graph in Figure 3–1 from a survey of 139 CEOs, presidents, and vice presidents of companies planning a CIM implementation. Note that 55 percent of the respondents listed *lack of in-house technical expertise* as a major obstacle to CIM implementation. To overcome this deterrent the education component for every employee must focus on several elements:

- The necessity for enterprise change to remain competitive in a national and world marketplace
- The need to support a new order in enterprise operations that includes team work, total quality, improved productivity, reduced waste, continuous improvement, common databases, and respect and consideration for all ideas regardless of the level from which they are initiated

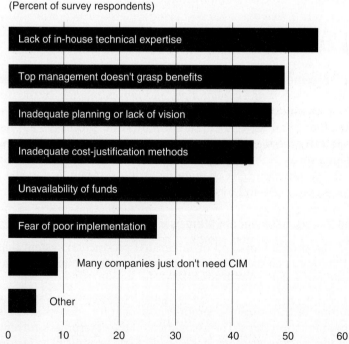

Figure 3–1   Obstacles to a CIM Implementation. (*Source:* Rehg, James A., *Computer Integrated Manufacturing.* Englewood Cliffs, NJ: Prentice Hall; 1994, p. 50)

■ The hardware and software necessary to implement a CIM system and the management strategy required to run the system successfully

CIM is not hardware and software; CIM is a way to manage the new technologies for improved market share and profitability. From the start of the implementation, all members of an organization must understand how CIM relates to their jobs; as a result, assessment and education must be first.

Step 2: Simplification—Elimination of Waste

In step 1 the enterprise was studied department by department, process by process, and activity by activity. In addition, a comprehensive effort to educate employees about the company, the need for change, and the technology planned for the future was put into place. From this knowledge base, step 2 is begun.

In most departments, CIM combines the manual operations and the islands of automation into an enterprise-wide, integrated solution. For example, the manual record-keeping process for tracking the location and quantity of inventory moves to a computerized system with all inventory information stored in a common computer database. If a company installs automation without first simplifying the process by eliminating the unneeded operations in the inventory process, it just automates many of the poor practices present in the manual system. In other situations, manual production operations yield to automated cells and production systems that permit greater productivity and improvement in basic performance measures. Building automated systems without eliminating operations that produce waste just automates the waste-production process. *Simplification* can be defined as follows:

> *Simplification is a process that removes waste from every operation or activity before that operation is implemented in the CIM solution.*

What is waste, and where is it found? Waste is every possible operation, move, or process that does not *add value* to the final product. If an activity log was made for raw material as it passed through manufacturing, the state of the material would be described as moving, waiting in queue, waiting for process setup, being processed, and being inspected. Only one of the five, *being processed* by the production machine, adds value to the part, and only if the production machine produces good parts. The remaining states are classified as *cost-added* operations because the part is *not* increasing in value as a result of the activity, and the enterprise continues to pay overhead cost during that time. In short, any operation or activity that does not add value to a product is cost-added and is waste. Removing all waste from operations is often not possible; however, all avoidable cost-added processes must be eliminated.

The search for waste often focuses on direct labor cost. Yet in most manufacturing operations direct labor usually accounts for only 2 to 10 percent of the cost of goods sold. Therefore, eliminating waste only from direct labor cost would not provide significant savings. However, case studies show that cost-added operations account for as much as 70 percent of all activities in the enterprise.

Eliminating as many of the cost-added operations as possible requires significant changes in the daily business and in the manufacturing process. The rules in Figure 3–2 offer a process designed to stimulate enterprise change and significantly reduce the level of cost-added operations. Armed with a comprehensive list of cost-added operations for an area, departments modify the processes so that half of the listed operations are eliminated.

Reducing waste in these significant amounts requires that the enterprise *attack waste fundamentally*. For example, a forklift was used originally to move material between production machines in a batch manufacturing operation. To reduce the transit time between machines—an identified cost-added activity—management links the production machines with a material-handling conveyor. Can you recognize the problem with this solution to waste reduction? Transit time is reduced and the process is quicker; however, material is still moved, work-in-process (WIP) inventory is not reduced, and the batch sizes have not changed significantly. The conveyor is a *superficial improvement*. A *fundamental improvement*, elimination of the transportation, is achieved by moving the production machines closer together. Identifying fundamental waste reduction requires that every employee be trained in the fundamentals of the business during the assessment step in the CIM implementation process.

The reduction of the waste by 50 percent in Figure 3–2 is the first step in the process to eliminate 90 percent of the identified cost-added processes. As the figure indicates, the 50 percent of the waste still remaining is reduced by 50 percent, so that the total waste present is only 25 percent of the original amount. The process continues until no more than 10 percent of the original waste is present. At this point, the manufacturing process is sufficiently stripped of unnecessary cost-added operations so that a successful CIM implementation is possible. The war on waste is never over, however, because employees use the *continuous improvement* technique to continue searching for ways to add value and to eliminate cost-added operations.

An aggressive simplification exercise on the shop floor to reduce cost-added activity assures that what gets automated by the CIM hardware and software is not bad production processes. When the simplification process is applied in every office and department in the enterprise, performance improves, and installed automation is effective.

Three-Step Rule for Eliminating Waste

Reduction	Total (%)
Reduce by 50%	50
Reduce by 50% again	75
Make it 10% of what is originally was	90

A world-class company works the three-step rule to arrive at a 90 percent reduction in waste.

Figure 3–2 Rules for Elimination of Cost-Added Operations.

Step 3: Implementation with Performance Measures

The first two steps prepared the enterprise for the CIM implementation; step 3 builds the system. Implementation with *performance measures* implies that the work cells and integrated systems are purchased and installed, but measurements of system performance must be included. In the past, companies made changes in the production system, such as installation of robot-automated work cells, but never measured the production performance gains provided by the automated cell. In addition, the efforts toward improvement in production often stopped with the installation of the automated system. Step 3 focuses on the design and implementation of a system based on the results of steps 1 and 2; however, the performance of the new system must be compared with performance from the past.

Implementation with performance measurement has two requirements: (1) measuring the success of the production process before the CIM implementation and at regular intervals afterward, and (2) recording the changes in key manufacturing and business parameters. The objective for the enterprise is to be competitive with the best in the world so that tracking success in world-class performance standards determines the level of improvement in the organization. The six performance standards frequently used for tracking progress include three standards shown in Figure 1–3 and three additional standards that measure design and production efficiency, employee productivity, and continuous improvement. The six key measurement parameters are described next:

1. *Product cycle time:* Product cycle time is the actual amount of time from the release of a manufacturing order to its final completion. For this parameter to be an effective measure of improvement, the setup time, queue time, move and transportation time, run time, and lot size must be included in the total. In some applications the total should also include the time required for completion of the product or part design.

2. *Inventory:* The inventory is measured as either material resident time (the time raw material or parts spend in manufacturing) or product velocity (the number of inventory turns by product). Most companies use the product velocity parameter.

3. *Setup times:* The setup time is a part of the product cycle time; however, improvement in setup is a key factor in the struggle to become competitive. As a result, most companies use setup time improvement as a measurement parameter.

4. *Quality:* Quality is either an order-qualifying or order-winning criterion for most products, so it is no surprise that improvement in this area would be a measurement parameter. Quality takes on many forms; however, the two areas used as primary measurement parameters are *first-time good parts* and *reduction of scrap and rework*. First-time good parts is a measure of how often the first part produced in a production run is within specifications. The reduction of scrap and rework is a measure of production

quality in two areas: scrap (unusable production that must be discarded) and rework (parts that are out of tolerance but that can be fixed with additional manufacturing operations).

5. *Employee output/productivity:* This term is a measure of the amount of output of goods and services per unit of input. Companies often use total employee hours worked and total units of output in this measurement parameter.

6. *Continuous improvement:* Of all the activities espoused in CIM, it is hard to find one more important than continuous improvement. As a result, some measure of the activity in this area is always used to judge the success of a CIM implementation. In most cases the measurement is just the number of improvement suggestions per employee per week or month.

Initially, all six parameters are measured and recorded to establish a baseline before starting implementation of any automation or process improvement. In enterprises with successful CIM programs, the performance measurements are reviewed monthly to track changes in the performance. During the *initial* CIM implementation, companies often start by measuring and tracking just two or three key parameters. They focus first on the parameter(s) that would provide the greatest initial payback. For example, inventory is often selected first because most manufacturing operations have excess inventory and the dollars saved through reductions are easy to identify. As control of the process is established, additional performance measures are tracked and improved.

The results listed in Figure 3–3 show the changes achieved over 18 months by a pump manufacturer with $25 million in sales. The remarkable improvements in *floor space* are a result of a shift from job shop (Figure 1–6b) and repetitive production (Figure 1–8) environments to a *structured product-flow* system (Figure 1–6c) which groups machines by product. At the start of the improvement process, the company had two choices: (1) add automation to the existing production system, or (2) look at production alternatives that would improve order-qualifying and order-winning criteria. If a decision had been made to automate the existing process—a job shop manufacturing strategy with large batches or lot sizes of pump housings—then such dramatic results would not have been achieved. Instead, the major sources of waste in the production process were identified and

Pump manufacturer's performance report card case history (pump housings)

Measurement parameters	Baseline	12 months	18 months
Cycle time	18 weeks	6 weeks	1 week
Inventory time	4	8	48
Scrap (percent of lot size)	13%	5%	0.06%
First time good parts	45%	85%	93%
Floor Space	1200 feet	500 feet	150 feet

Figure 3–3   Results of Improved Business Operations.

attacked with fundamental changes in the way pump housings were manufactured. As a result, the automation was significantly more effective in improving productivity than would have been possible if the same automation had been added to the old production system. The investment for either approach would be about the same. However, the return on the investment in terms of competitiveness in the marketplace was much higher for the second choice because the company chose to adopt a new production strategy. The success achieved was a result of hard work in the assessment and simplification phase of the CIM system design. Compare the 18-month results with the standards listed in Figure 1–3.

The final step in the CIM implementation process includes the acquisition and installation of the hardware and software to the specifications developed in steps 1 and 2. When assessment and simplification are completed, and a performance measurement system is in place, then a successful implementation of hardware and software is assured. Unfortunately, many companies implement CIM automation beginning with step 3. As a result, they automate all the disorder of the poor production processes and produce waste at a record rate.

3-3 MAKING THE PROCESS WORK

The three-step process identifies three major tasks that must be performed for a successful CIM implementation. The specific process used in each step is a function of a number of variables. For example, the process used by the pump manufacturer featured in Figure 3–3 was affected by the type of manufacturing systems used; the fact that management and labor had a good working relationship; current competitive position in the marketplace; earlier work in identifying training needs across the company; and experience gained in a small automation project completed previously. The specific process used by companies as they address the implementation of CIM principles will vary depending on the conditions present and the corporate culture. Some of the elements found in successful CIM implementation processes include these:

- Use of a program name different from computer-integrated manufacturing
- Support for the program beginning with the CEO
- Use of multifunctional employee teams at all levels in the process
- A willingness to look at all processes and products for potential productivity gains
- A willingness to accept recommendations for process and product improvement from every employee
- A willingness to accept a 3- to 5-year payback time for the investment.

Most companies choose not to identify their project with the CIM acronym because they want a title that is unique to their operations. For example, the pump manufacturer in Figure 3–3 named its project the *Rapid Customer Response System* *(RCRS)*. The acronym avoided the use of the term *CIM* in the project title because the firm believed that the title should communicate the goal of the project to every-one in the company. Nevertheless, a study of the RCRS process indicates that it follows the philosophy and elements defined by CIM.

Several conditions guarantee the failure of a CIM implementation from the start, and lukewarm support from the top management is at the top of the list. The corporation has to change when CIM is implemented, and change will not occur without support of top management. Other critical elements include the willingness to form teams with representatives from all corporate departments and all employee levels. Equally important is a belief that everyone in the organization has information and suggestions that are critical to a successful CIM implementation.

Most enterprises implementing CIM give the project a unique name and approach the assessment and simplification phase with a technique that fits their unique operations, products, and corporate culture. Regardless of the project name or the technique used to study the corporate operations, the end result is a CIM-based enterprise that is better equipped to compete in global markets.

3-4 AUTOMATED PRODUCTION

Reengineering of the manufacturing operation frequently requires the use of automation to improve production performance and to meet order-qualifying and order-winning criteria. The degree of automation is a function of manufacturing and assembly specifications, labor conditions, and competitive pressure. For example, some production operations and assembly tasks are performed better by human operators, whereas others require the precision, speed, and repeatability of automated machines. Labor costs and work requirements play a role in the degree of automation as well. Automation is necessary on some products to meet the competitive price demanded by the marketplace. In other situations, if human operators are used, the required work motions could cause a health hazard (e.g., carpal tunnel syndrome). In general, manufacturing is using more automation to meet the performance measures demanded by the competition.

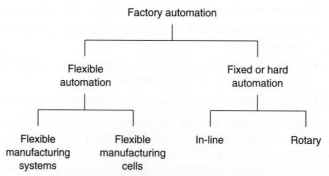

Figure 3–4 Types of Factory Automation.

The types of automation common in manufacturing are classified in Figure 3–4. The relationship between production capacity, production flexibility, manufacturing systems, and type of automation is illustrated in Figure 3–5. Study the two figures until the automation types and relationships are clear. Note that robots are frequently used in all manufacturing systems except *project*, where only limited applications occur. Most robot applications occur in either flexible or fixed automation cells and systems, so a description of these areas is a good starting point.

3-5 FLEXIBLE AUTOMATION

Flexible automation enables flexible manufacturing. Flexibility is often an order-qualifying or order-winning criterion and is a key performance measure listed in Figure 1–3. In manufacturing flexibility is often described as (1) the ability to adapt to engineering changes in the part; (2) the increase in the number of similar parts produced on the system; (3) the ability to accommodate routing changes that allow a part to be produced on different types of machines; and (4) the ability to rapidly change the setup on the same machine or system from one type of production to another. In each case, the critical point is that properly designed production areas

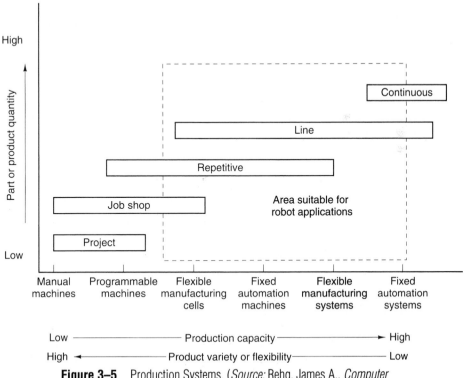

Figure 3–5 Production Systems. (*Source:* Rehg, James A., *Computer Integrated Manufacturing.* Englewood Cliffs, NJ: Prentice Hall; 1994, p. 294)

make production planning and scheduling easier and result in quicker response to customer needs. Flexible manufacturing takes two forms: the *flexible manufacturing system (FMS)* and the *flexible manufacturing cell (FMC)*.

An FMS is defined in the *Automation Encyclopedia* (Graham, 1988) as follows:

A flexible manufacturing system is one manufacturing machine or multiple machines which are integrated by an automated material handling system, whose operation is managed by a computerized control system.

The definition implies features such as numerical control (NC) or smart production machine tools, automatic material handling, central computer control, production of a variety of parts in random order, and data collection and integration. In general, an FMS is an integrated collection of production hardware linked together by computer software. The highly integrated nature of the FMS requires the design and implementation of the system as a unit.

An FMC is defined in the *Automation Encyclopedia* (Graham, 1988) as follows:

An FMC is a group of related machines which perform a particular process or step in a larger manufacturing process.

Based on this definition, an FMC is a group of integrated product machines that perform part of a larger process. The FMC could be viewed as the production building blocks used to assemble an FMS. Figure 3–5 puts FMS applications and FMCs into a manufacturing perspective. Note that a single FMC would be used for medium-capacity production in job shop and repetitive manufacturing systems, whereas an FMS would more commonly be used to build a line-type manufacturing system. However, it is often difficult to differentiate between a small FMS and a large FMC.

For the purposes of this text, the study of flexible manufacturing will focus on two configurations of FMCs: (1) the FMC that is used as an integral part of a larger FMS, and (2) the FMC that results from automating a manual production machine or area. To obtain a clearer picture of the two configurations of FMCs, consider the Application Notes beginning on p. 89.

Both case studies describe similar FMCs used to assemble electronic circuit boards with either surface-mount devices or with conventional electronic components. In both cells, vision was used to check part location; multiple grippers were used to pick a wide variety of part geometries; and the robot was the primary production machine in the work cell. The two case studies exhibit one significant difference.

In the Technophone case study the FMC was a part of a larger FMS, with the cellular circuit boards and the production system designed simultaneously. The product was designed to be manufactured by the FMS, which was built for maximum compatibility with the product. In the TRT Thompson case study a manually produced product was converted to automated assembly. The FMC was not a part of a larger automated line, and the automation hardware had to work with a product designed for manual assembly. The design process for the FMC in each case was similar; however, adding automation to a previously manual production system creates some unique problems. A process to automate manual production cells

APPLICATION NOTE
ELECTRONIC ASSEMBLY OF ODD-FORM SURFACE-MOUNT DEVICES*

Designed by John Brown Automation, the electronic assembly system at Technophone (Figure 3–6) operates 24 hours a day assembling boards that are used in mobile cellular telephones.

The electronic assembly system is a vital part of an integrated production line that makes 8000 telephones per month.

By automating its production line, Technophone has not only been able to han-

Figure 3–6 Assembly of Cellular Phone Circuit Boards. (Courtesy of Adept Technology, Inc.)

dle the demand for cellular phones, but it has also reduced labor costs and improved product quality.

System Configuration

Fast, precise printed circuit board assembly is accomplished by an AdeptOne robot with integrated vision. The AdeptOne assembles all of the odd components at a single station, at the rate of one component every four seconds.

With its tool-changing capability, the robot can assemble ten devices on the two different Technophone boards: seven different connectors, a spacing washer, a filter and a coil. The robot has four grippers to choose from; two of the grippers use vacuum to pick up components. The other two are mechanical grippers—one with a flat jaw and the other with a curved jaw.

An AdeptVision XGS-II grayscale vision system is integrated with the robot system.

The vision system's unique recognition algorithms ensure component placement accuracy to ±0.2 mm.

Adept's patented AIM (Assembly Information Manager) software manages the printed circuit board assembly data—simplifying cell development and operation.

Workcell Operation

The electronic assembly workcell (Figure 3–7) is one station in a larger production line. Before a board arrives at the Adept station, it has already passed through a screen printer, a solder paste-applicator and an automatic SMD placement machine.

The workcell builds boards in batches of 200-500, 24 hours a day. Two operators run the entire line—replenishing component supplies, handling batch changeover and dealing with complications. All of the assembly programs are held in the robot controller so batch change-over requires just one simple instruction.

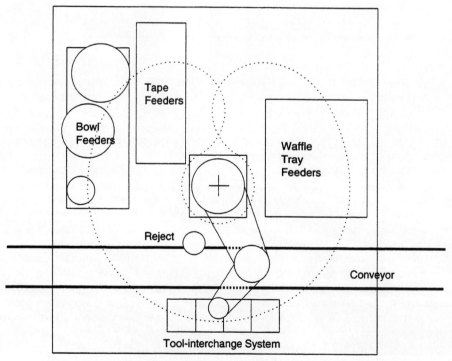

Figure 3–7 Layout of Technophone Flexible Manufacturing Cell. (Courtesy of Adept Technology, Inc.)

Prior to assembly, the vision system verifies the board type. A camera mounted on the robot arm identifies four fiducial marks on each board, and the robot controller adapts the assembly program for the specific board type, location and orientation. A collection of tape-feed devices, vibratory bowl feeders and waffle trays present the various components to the robot in the correct orientation. After the robot picks up a component, a second camera positioned just below the assembly area inspects the component.

If the part passes the inspection, assembly continues. However, if it is defective—for example, if it is the wrong part or if its leads are skewed—the robot drops it down a reject chute and selects a new component.

Including gripper changeover time, the robot achieves cycle times of 32 seconds for a 7-component board and 18 seconds for a 3-component board.

The system at Technophone can produce 900 parts per hour and improved manufacturing performance in both quality and cost. The investment in the system was paid back in two and one half years.

*(Reprinted with permission of Adept Technology, Inc.)

APPLICATION NOTE
THROUGH-HOLE PRINTED CIRCUIT BOARD ASSEMBLY*

To maintain a competitive edge in the marketplace, French electronics manufacturer TRT Thomson needed to replace its labor-intensive board assembly operations with a solution offering higher productivity and better quality.

TRT required a solution that was flexible enough to handle a variety of components and several different board types—as TRT produced volumes as small as 10 boards per month of each design.

The company considered hard automation, but this solution would have required three different stations: one assembling axial components, a second assembling radial components and a third assembling dual in-line packages (DIPs). Moreover, this solution would have required board-handling between machines.

In the end, TRT Thomson opted to replace its manual production with a flexible automation solution designed by French system integrator SCEMI.

System Configuration

The installation at TRT performs both component preparation and insertion at a single station. An AdeptOne robot with integrated vision works three shifts per day stuffing printed circuit boards at a rate of 800 components per hour.

There are three different types of components to be assembled: axial, radial and DIPs. In the cell's feeding equipment there is enough space for up to 50 reels of axial components, two of radial components and 60 tubes of dual in-line packages.

In total, the robot has to choose among 112 different components. With a compliant wrist and an automatic gripper-change system, the AdeptOne can easily handle the number and variety of components—quickly and accurately.

Workcell Operation

An automated handling system loads the boards onto the robotic assembly station. A bar code on the side of the board communicates to the robot which assembly program the robot must access.

An AdeptVision system verifies the location of a printed circuit board by "looking" at fiducial marks on the face of the board. An Adept MC robot controller then makes the necessary modifications to the assembly sequence.

With the help of an Adept vision system, the robot prepares and inserts the components into the board. If a component's leads are difficult to insert, the robot will use vision to "look" at the holes the leads should pass through. The robot controller will then modify the assembly program as necessary.

To maintain accuracy over time, the vision system is used to recalibrate the robot's working envelope to within 0.02 of a micrometer.

Once a board has been assembled, it is automatically unloaded from the robotic assembly station.

The station at TRT Thomson requires minimal supervision—operators need only ensure that the component feeders are full and available to sort out any issues. The system increased productivity and quality, and provided a four-year payback.

*(Reprinted with permission of Adept Technology, Inc.)

Figure 3–8   Assembly of Printer Circuit Boards. (Courtesy of Adept Technology, Inc.)

and a discussion of the problems encountered is the focus of an in-depth case study introduced later in the chapter.

3-6 FIXED AUTOMATION

A fixed automation manufacturing system able to produce a large volume of parts or products is often necessary to meet market demands. An example of this type of product is the disposable razor or lighter. In the automotive industry fixed automation is used to produce parts like fuel pumps and tires. Automation in this category has the following characteristics:

- A series of closely spaced production stations linked by material-handling devices to move the parts from one machine to the next
- A sequential production process with each station performing one of the process steps, and the cycle time at each machine usually about equal
- A system in which raw material and components are introduced at various points throughout the system, and finished parts or products exit at the end
- The number of stations in the system dictated by the complexity of the production process implemented
- A system that uses high-speed robots
- A tightly linked production system that produces either one or a few different products
- Either fast servo robots (Figures 2–5 and 2–18) or pneumatic robots (Figures 2–4 and 2–11).

Fixed automation systems are implemented in two configurations: *in-line* and *rotary*.

In-Line Fixed Automation

An *in-line fixed automation* system (Figure 1–7) consists of a linear work flow through a series of workstations. Note that the system in the figure is designed to produce two different small products. These two products use the same machines for their first few production operations and then diverge into separate manufacturing paths. In this example, the production system uses thirty-seven pneumatic robots to move the product and parts between assembly machines and production stations.

Although most fixed automation systems are highly automated, it is possible to include manual stations along the line to handle operations that cannot be automated economically.

Rotary-Type Fixed Automation

Rotary-type fixed automation locates the production stations around a circular table or dial. As a result, this type of production system is called an *indexing machine* or a *dial index machine*. The system pictured in Figure 2–23 has twelve production stations positioned on the rotary table and uses seven pneumatic robots to support the assembly. In this rotary system and in the in-line system in Figure 1–7, automatic parts feeders supply the pneumatic robots with components in the sequential assembly process. Such a system is limited to smaller workpieces and has fewer production stations than the in-line type.

3-7 IMPLEMENTING AUTOMATED WORK CELLS

The implementation of automated work cells follows a process similar to that defined for CIM. Two major issues must be addressed early in the design process: (1) the use of current production hardware versus the purchase of new process machines, and (2) the adoption of fixed versus flexible automation. An understanding of the design process starts with a look at the first issue.

New versus Existing Production Machines

Work-cell automation projects fall into two categories: *new automation* and *existing production cells*. In both types of projects, the automation can focus on a single production work cell or on the automation and integration of a group of cells. In the first category, new automation, both the production process machinery and the automation hardware are specified in the automation project.

In the second category, existing production cells, the current production machinery is integrated with new automation hardware and software to upgrade the production process. The decision to choose existing production machines over new ones is based on *market conditions* and *economics*. For example, if the order-winning criteria require a level of quality or price that current manual production machines cannot meet, then construction of a new automated production system is necessary to be competitive. If the cost of a new system prevents that course of action, however, then the company has two choices: (1) discontinue the product and abandon that market area, or (2) position the product in a niche in the larger market. In the second case, the product could be designed to fill a special niche where quality is not as critical. With the quality issue satisfied, the current production equipment could be integrated and automated so that the cost is competitive in the niche area. In the pump manufacturing example (Figure 3–3), the process machines had the necessary precision but were not delivering order-winning values for cost, production flexibility, or cycle time due to poor work flow. Reorganizing the production machines and adding the appropriate automation to the reconfigured work cells was a cost-effective solution. When a new automation cell is designed and built with new process machines and automation hardware and software, the integration requirements for all of the equipment are specified in the purchase contract. As a result, the implementation of new automation cells often presents fewer problems.

Fixed versus Flexible Automation

In the second issue, the use of fixed versus flexible automation is examined. Parts and products manufactured using fixed automation generally have the following characteristics:

- *Low product variability:* The product design is well defined, and no major changes are planned in size, shape, part count, and material. The production of electrical wall switches for houses is a good example of a product with low variability. The switch design is well tested, and the likelihood for a change in the design is very small.
- *Predictable demand:* The demand for the part or product is stable for a 2- to 5-year time frame. For example, the automotive industry frequently signs multiyear contracts with suppliers of parts like horns and water pumps. With this multiyear contract commitment, an investment in a fixed automation system is justified.
- *High production rates:* The number of units produced per hour or day is usually high. Consider the production rates necessary to match the number of disposable razors sold per day around the world. High-speed fixed automation is the logical choice for such a production system.
- *Cost pressures:* The prices of many products, like disposable razors, are set by market conditions, and the manufacturer must meet the market price to be competitive. Frequently, fixed automation is the system of choice when high volume and low cost must be achieved.

When one or more of these product characteristics are present, fixed automation is considered for the production system.

The Through-Hole Printed Circuit Board Assembly case study describes a rationale for the selection of flexible automation over a fixed system. TRT Thomson chose a flexible system because the variation in components would require three different fixed automation work cells linked by material handling. With the cost and production speed of the fixed and flexible systems almost equal, the flexible system was chosen because it offered the opportunity for *future cost avoidance.* Fixed manufactured systems are often scrapped when the products produced on them are no longer sold. In contrast, flexible manufacturing cells and systems are retooled to produce a different product with a minimum investment in new hardware. In general, flexible automation is used in these situations:

- The product mix requires a combination of different parts and products to be manufactured from the same production system.
- The product family has a history of unavoidable engineering changes that alter production requirements.
- The product or part family will expand with similar but not identical models.
- Production volumes are moderate, and demand is not as predictable.

CIM is not just the application of automation hardware and software; however, the full benefits of CIM require the judicious use of automation. Therefore, work cells with a wide variety of automation systems, including robotics, will be part of most manufacturing integration projects. Many cases of poorly designed systems indicate the need for a good design process. The design, operation, and troubleshooting of work cells using robot automation is addressed next.

3-8 AUTOMATION CASE STUDY

The case study introduced in this section continues through the remaining chapters of this book. The case provides a detailed look at the process used to implement work-cell automation within a computer-integrated manufacturing plan. The problem presented in the case, although fictional, is derived from actual industry data.

CASE: CIM AUTOMATION AT WEST-ELECTRIC

PART 1: INTRODUCTION

On the drive to work, Bill Baxter usually reviewed the work scheduled for the day, but today he had a hard time concentrating. It had been about a year since he joined West-Electric (W-E), and much of the time was spent working through the initial phase of the Customer-Driven Manufacturing (CDM) project, West-Electric's response to CIM. If implemented fully, the plan would make significant changes in the division's manufacturing operations. Bill was ready to move from the planning to the implementation stage, and if the rumors were true, the meeting at 10 this morning would begin the automation phase.

It was 9, and Bill had just enough time before the meeting to review the assessment data his team had pulled together in the past several months. They had spent about 2 days a week for 6 months working on benchmarks for a number of performance measures on the division's major product, jet engine and gas turbine blades. The process had been developed at the corporate level, but plant manager Roger Walker allowed them to take some liberties whenever it benefited the assessment process. Roger was the best plant manager with whom Bill had been associated. Rumor had it that his plant would get the first automation project because of his reputation across the company.

"Good morning, Bill . . . I need a few minutes of your time."

Bill looked up to see Roger standing in the doorway to his office. "Sure, come on in. I was just reviewing the CDM project report for the meeting this morning."

"That meeting is what I want to talk to you about. Bill, I'm naming you the project leader for the first automation projects to support the CDM initiative. I will announce it in the morning meeting. You look a little surprised."

"Well, I certainly hoped to be part of the project team, but I thought the leadership would go to a more senior engineer in manufacturing."

"You're my choice for several reasons. You've had experience in installing automation in your previous job, plus you've had the most recent academic exposure to the automation technology we need to use," Roger said. "I also want a fresh look at how we make parts and want someone without any emotional attachment to the current processes we use. The project will focus on automation of the turbine blade manufacturing line." After a few moments, Roger asked, "Are you OK with this assignment?"

"It's a great opportunity . . . you'll get my best effort," Bill responded.

Laying a folder on Bill's desk, Roger said, "I asked Ellen Becker in Human Resources to put together a short overview of each of the team members so you know the type of support you have available. Hold any other questions until after the meeting. See you shortly."

Bill indicated he understood and picked up the overview of his team.

West-Electric Automation Team

Bill Baxter (Manufacturing Engineer).   One year with W-E following the completion of a BS and MS in mechanical engineering. Two years of co-op experience with another division of W-E during graduate study. Previous experience includes 5 years installing and troubleshooting automated manufacturing systems following the completion of an associate of science degree (AS) in mechanical engineering technology.

Ellen Becker (Director of Human Resources).   Ten years with W-E with a master's degree in human resources earned while working for the company.

Marci Hatcher (Manufacturing Engineer).   Seven years with W-E with the most recent work on process improvement and standards for an MRP II project. Previous experience includes 6 years in methods and work simplification in a metal fabricating company. Educational background: BS in industrial engineering and MBA.

Ted Holcome (Design Drafter).   Fifteen years with W-E in a broad range of documentation and design projects. Education includes an AS in engineering graphics and recent update training in three-dimensional and solid modeling CAD, plus training on a new work-cell design and simulation software package.

Mike Perry (Senior Designer).   Eleven years with W-E working on the design of production machines and fixtures. A BS in mechanical engineering and a good background in CAD and finite-element analysis.

Jerry Thompson (Senior Electrical Technician).   Eight years with W-E with experience in installation of new equipment and the maintenance of current systems. Education includes an AS in industrial electronics technology and recent training to install and program the programmable logic controller, which was picked as the standard for W-E.

The First Meeting

Bill got to the meeting early so that he could be there as everyone else arrived. He knew Marci and Ellen from the initial CDM project and had worked through a production fixture design problem with Mike shortly after joining the company.

He didn't know the other team members and wanted to meet them as they arrived. At exactly 10, Roger Walker entered the conference room, motioned for quiet, glanced at his notes, and began to speak.

This week I will start three productivity projects as a result of the work done in the CDM initiative. The one in the product design area focuses on a defect-free design process by developing a design for manufacturing and assembly process with a six-sigma design component (a technique that widens allowable part variation to ensure successful part assembly). The second project in production planning and control will work on alleviating some quality and inventory issues. Your project will add automation to the turbine blade manufacturing line. All of you participated to some degree in the CDM process. Bill, Marci, and Ellen were on the enterprise team that established the process and completed the assessment of our division in the areas of *technology*, *human resources*, and *process systems*. I'm not going to cover the initial CDM results in detail; however, a review of some findings will be a good introduction to the project.

A study of our *current manufacturing technology* indicated that we were current with the technology used to achieve process-specific results. But no technology was in place to integrate and automate the processes or to make manufacturing data available to enterprise units on a real-time basis.

In the *human resources* area the results were mixed. In general, the staff had the capability to implement and operate the technology required to meet internal and external customer expectations. The option to use consultants or to buy the automation from outside companies was considered. It was rejected, however, because I know we have the talent to do the job, and I want to grow that capability for future automation projects. Though we had the talent, we had not done a good job of informing everyone in the division of the need for change in design, production control, and manufacturing. This shortcoming was especially apparent in the organized labor component in manufacturing. As you know, the case for a stronger customer focus has been the theme of my communications to your departments for the past year. In addition, Ellen has done a good job of getting that message to the production areas through the work-cell, continuous-improvement team meetings.

Our ISO 9000 project contributed much of the data for assessment of the *process systems* area. The ISO documentation gave us some insight into how the current production processes operate. When it became apparent that the turbine blade line was targeted for the first manufacturing automation project, additional assessment was performed. All of that documentation is available to you for this project.

You may be interested in the rationale for selecting the turbine blade line for the initial project. Plans are in the works to build another gas turbine assembly plant in about 30 months, including a blade manufacturing line. The design for the new line calls for tighter integration between manufacturing operations and a significant increase in automation. Automating our current blade line will serve two purposes: our productivity and capacity will be increased, and we will learn how to effectively integrate and automate the production of blades.

As you develop the plan for the automation of the line, consider the following: We are the first division in the company to get capital dollars for new automation. Second, the level of future funding is always affected by the success of current projects. Third, what we do here will be closely watched by the folks at the corporate level. What I'm saying is we need a successful project. Do everything reasonable to guarantee that the design on paper will perform when it's built.

I have asked Bill Baxter to lead the project work. Ellen has been assigned to all three automation teams, so her support is divided accordingly. Your supervisors recommended you for the project and are aware of my interest in its success. While Bill will work full-time on the project, the rest of the team will have some other department work. I made it clear to your supervisors that this project takes priority.

One final comment before I turn the meeting over to Bill. The project must have a 2-year or less payback, and the completion date for the automation of the line must be less than 18 months. That's about it from my end. You all know that I believe in the CDM concepts and know that this project can be successful. Bill, is there anything you want to add?

Bill Baxter leaned back in his chair, looked slowly around the room, and addressed the group.

I would like to get started on the project as soon as possible. I have some background material for each of you that includes an overview of the blade line and a description of the CDM design process we will follow. Please study the material and meet back here tomorrow morning at 9.

PART 2:   TURBINE BLADE PRODUCTION

The turbine blade production line manufactures 70,000 blades per week in seven sizes; the production is about equally divided over the different geometries. The shape of the blades is similar (Figure 3–9), but the weight varies from 3 to 16 ounces. The blades are manufactured from seven different lengths of titanium or stainless steel bar stock that varies in diameter from 3/4 to 1 1/2 inches. The current production system uses manual production machines operated on three shifts for 5 days a week. Saturday overtime is often necessary to meet delivery schedules when the production capacity is exceeded or when equipment problems disrupt normal production flow.

The blade production uses a six-step production process for five blade shapes and a four-step process for the other two blades. During the six-step blade manufacturing process, the raw material passes through the following five stages

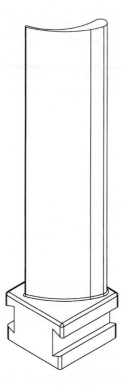

Figure 3–9 Forged Turbine Blade.

(Figure 3–10): *slug, extrusion, upset, block forge, final forge and trim.* In the four-step process the *extrusion* and *upset* operations are eliminated because the blades are small. The process flow for all of the products, illustrated in Figure 3–11, is divided into three major operations: slug preparation, airfoil forging, and final production finishing. Detailed process flowcharts for each of the major operations are included with the following descriptions.

Step 1: Raw material inspection.   Production begins (Figure 3–12) with the selection of a 20-foot titanium or stainless steel bar with a diameter of 0.75, 1.0, 1.125, 1.25, or 1.5 inches. An operator passes the bar through an ultrasonic inspection station that checks the rod for internal flaws in the metal, such as voids or impurities. A visual and audio output indicates a bad section of bar that cannot be used in the subsequent forging processes. The bad section is located by slowly rotating the bar in the machine until the exact location is found. The spot is marked by the operator and eliminated when the bar is cut into slugs in the next operation.

Step 2: Production of slugs.   Good bar sections are fed into an automatic shearing machine (Figure 3–12) where induction heaters raise the bar temperature to 1300 degrees Fahrenheit (1300°F) in preparation for the shearing operation. The shear cuts the bar to a preset length, and the slug drops into a tote pan for transport to the next operation.

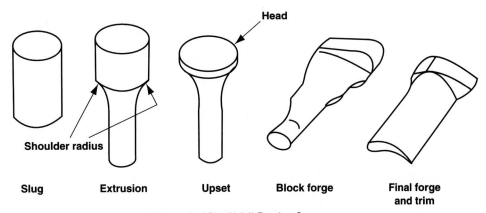

Figure 3–10 Airfoil Forging Sequence.

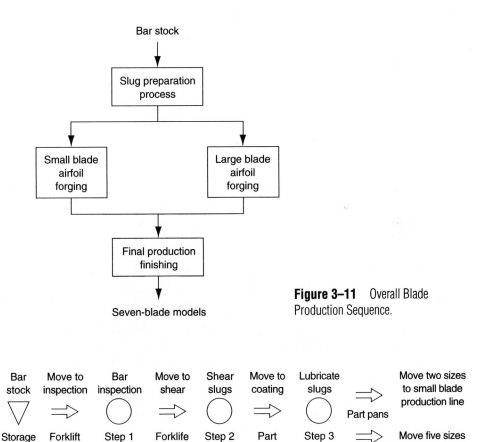

Figure 3–11 Overall Blade Production Sequence.

Figure 3–12 Slug Preparation Process.

Step 3: Slug lubrication. The slug lubrication area (Figure 3–12) coats the slugs with a ceramic silica glass lubricant that enhances the forging operation. The lubricant is applied using three manual spray booths. The operators pick up a tote of approximately 500 slugs from the shear or from inventory and coat them in batches of 72 or 90, depending on size. The slugs are arranged in a nine-by-eight or a ten-by-nine array (Figure 3–13) on a rotating platform. The operator picks up the spray gun from a hook on the booth and applies the lubricant as the table is manually rotated to obtain an even coat on all visible surfaces. After a 10-minute drying time, the slugs are removed and placed in a tote for delivery to the next process. The larger size slugs go to the large blade production process (Figure 3–14), and the two smaller sizes go to the small blade production process.

Step 4 Extrusion. Three manual work cells, each equipped with an oven and a 500-ton capacity forge, produce 3000 to 3600 extrusions per shift for three shifts. The slugs are placed on a rotating circular table inside the oven and heated to 1800°F. It takes about 500 seconds for the larger diameter slugs to reach saturation temperature; the smaller parts require only 400 seconds. At the start of a shift, slugs are loaded onto the oven's indexing table at 15-second intervals until the first loaded slug has been in the oven for the required time. At that point, the operator picks up a cold-coated slug from the tote basket with 20-inch tongs and opens the automatic door of the oven with a foot switch. The cold slug is placed on the rotating shelf in an upright position, and a hot slug is removed for the extrusion process. The hot slug is dropped into the extrusion die, and the forge is cycled by the operator using two start buttons. After the 3-second extrusion process, the operator waits until a 15-second timer indicates that the extrusion has cooled sufficiently to

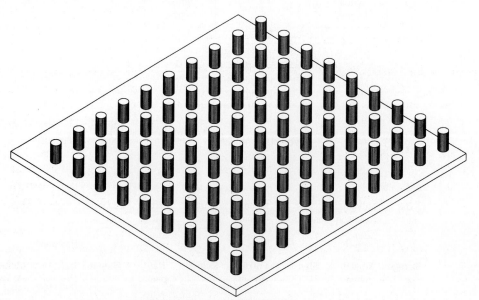

Figure 3–13 Lubrication Spray Table.

be removed from the extrusion die. The operator uses a foot switch to activate an ejector, which pushes the extrusion out of the die so that the operator can transfer it with tongs to a finished parts tote. The operator uses a manual sprayer hanging on the forge to coat the upper and lower halves of the extrusion dies with graphite lubricant after each forge cycle. The total cycle time for the process is 30 seconds. Seventy-five to 300 tons of pressure are required, depending on the material and size of the slug, and a quick visual inspection of the surface finish is performed as the extrusion is removed from the forge. The operator works in a hot environment and must wear gloves for protection. In addition to the visual inspection, the shoulder radius (Figure 3–10) is measured with gages and with an optical comparator at 200-part intervals.

Step 5: Upset forging.   The upset forging operation (Figure 3–14) forms the head of the bar (Figure 3–10) in preparation for the block forge operation. This process does not require a ceramic silica glass lubrication coat before forging. The three manual work cells are nearly identical in layout and operation to the extrusion process with an oven and a 100-ton press present. The production rates are the same as those for the extrusion process; however, the cycle time is shorter (15 seconds) because the die does not require the graphite lubricant application between cycles and the forge cycles in 1.5 seconds. A cold part must be in the oven 450 seconds to reach the correct temperature; as a result, the oven must be loaded with 30 parts (a part every 15 seconds for 450 seconds) at the start of a shift. After the initial loading, an audible signal every 15 seconds indicates that the parts shelf in the oven has indexed, and a part is at 1800°F, ready for upset forging. After the upset operation the part is immediately removed from the die and dropped randomly into a finished part basket. The production requires only five to 30 tons of the rated forge capacity for the five different part sizes manufactured. No visual inspection is performed on parts as they are produced, but the radius (Figure 3–10) of the upset is measured with a dial indicator at 100-part intervals.

Step 6: Block forge.   Block forge gives the airfoil the rough geometry needed for a successful finish-forge operation. The three block-forge work cells have the same general layout and operation as the two previous steps, extrusion and upset forging. The ceramic silica glass lubrication is applied to the upset part using either a hand-dip method for larger parts or an automated electrostatic spray line for the smaller parts. In both cases, the parts are manually placed in carriers that prevent contact between coated parts. The oven-loading sequence, temperature saturation time, and process cycle time are identical to the upset operation. However, the dies have to receive a graphite lubrication spray between forge cycles. No inspection is performed on the block-forged parts.

Step 7: Final forge and trim.   The four-cell final forging operation (Figure 3–14) has an equipment layout and process similar to the previous three processes. The primary exception is the addition of a trim press used to remove excess material after the final forge operation. The cycle time is 30 seconds (equally split between forging and trimming), and the time to reach forging temperature is 900 seconds

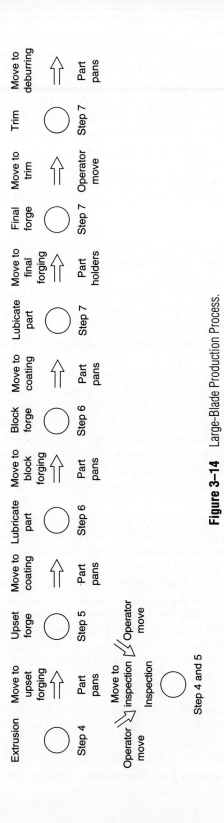

Figure 3–14 Large-Blade Production Process.

for the largest part and 240 seconds for the smallest. As a result, the number of parts in the oven is a function of part size. The coating of the block-forged part with ceramic silica glass lubricant and the graphite spray for the upper and lower dies must be applied carefully to produce parts within design tolerances. The operator never releases the part from the tongs as the forge cycles. The round shaft at the end of the block-forged part (Figure 3–10) is used to hold the part throughout the process. After final forging, the operator cools the larger end of the airfoil by dipping it into water for 2 seconds, then places the blade in a trim press. The trim press removes the extra metal on the finished part, called flashing, that came out between the upper and lower die faces during final forging. At this point the airfoil has the desired shape.

Step 8: Alternate process for small airfoils.   The smaller blades skip the extrusion and upset operations and go from lubrication of slugs to block forging. The process in block forging (Figure 3–15) of the small parts has several variations. The oven is replaced with two induction heaters, and the forging operation is performed with an air hammer or drop forge. The operator has a 10-inch tongs in the right hand to remove the hot slug and holds a cold slug in the left hand. The operator removes a hot slug from one of the induction heaters using the tongs and immediately drops in the cold slug with the left hand. The block forge is performed on the slug, and the part is placed in a finished tote. The process is repeated with a slug from the other induction heater. The cycle time for the process is about two parts every 15 seconds.

The final forge on smaller parts also uses two induction heaters and an air hammer. After the final forge, however, a trimming process for removal of flash is performed. The cycle time for the final forging of one airfoil is the same as that for the block-forge operation—7.5 seconds; however, the trimming operation adds another 10 seconds to the single-part cycle time.

Step 9: Deburring.   The final production finishing process (Figure 3–16) starts with deburring. Rough edges caused by the trim press operation and other irregularities on the airfoil surface are removed in this manual operation.

Step 10: Heat-treating.   The blades are heat-treated in a vacuum oven to remove the stresses developed in the forging operations. The total time for this operation is 8 hours and includes manually loading 400 parts into the oven, bringing the oven up to temperature, heat-treating at specified temperature, and controlled cooling.

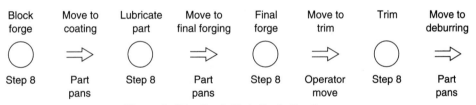

Figure 3–15   Small-Blade Production Process.

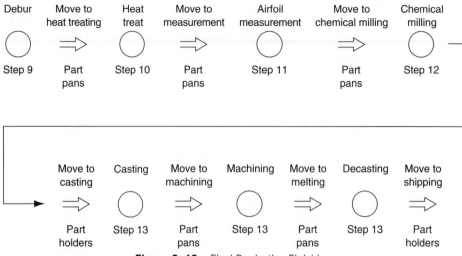

Figure 3–16 Final Production Finishing.

Step 11: Measuring airfoil cross section.   The complex shape of the airfoil is checked by a custom inspection machine (Figure 3–16) that simultaneously measures a minimum of three cross sections of the blade. The operator removes a finished blade with one hand and places it in a finished part tote basket. A new part is picked up from a container with the other hand and placed in the measurement fixture. The fixture is manually closed, and the measurements are recorded in the system memory and displayed on the machine front panel. The nominal value for the cross section at this step is 10 thousandth oversize. Oversized parts are marked and divided into five oversize ranges, from 1 thousandth to 13 thousandth in 3-thousandth increments. Any blade over 13 requires rework, and any blade under 1 is scrapped. The measurement data are also displayed on a system monitor to support the statistical process control (SPC) analysis of the manufacturing process. The cycle time for this step is 17 seconds.

Step 12: Chemical milling.   The surface of the airfoil is chemically milled to remove the oversized stock and to produce the finish required for operation. Based on measured values, blades are manually placed in special baskets and tumbled in hydrofluoric and chromic acid for varying lengths of time. The milling time varies from 4 minutes to more than 15 minutes.

Step 13: Machining base.   The complex shape of the part (Figure 3–9) makes holding the part while machining the base difficult. To simplify the fixture problem during base machining operations, the blade is encased in a metal alloy with a melting point of 475°F. After the casting process, the entire blade has the shape of a square bar with flat ends. The casting machine has three molds arranged at equal distances around a 4-foot diameter indexing table. One mold is always in the blade load or unload position; the second is positioned under a device that directs a metered amount of the molten alloy into the mold; and the third is in a cooling position. The operator unloads a finished part, loads a new blade into the mold, presses a switch to index the table one-third of a revolution, and repeats the process. After the machining of the base, the alloy is removed by melting and the finished blade is packaged for shipment.

Additional data on the blade production line.

- The average labor rate on the line is nine dollars per hour plus 42 percent for fringe benefits, with overtime paid at time and one-half.
- The 480-minute production time per shift for the work cells is reduced to 400 minutes due to die and tooling changes, pre-heating dies, machine maintenance, and initial loading of ovens.
- Extra floating operators are necessary to relieve regular cell operators for scheduled breaks.
- Tooling in each cluster of work cells is interchangeable.
- Each electric-heat oven has a door opening measuring 18 inches high by 15 inches wide, and the distance from the door to the parts is 16 inches.
- The die holders are normally 24 inches square, and the distance from the front of the die to the part cavity is normally 18 inches.
- Work in the extrusion, upset, block, and final forge areas is both physically and mentally stressful due to heat from the frequent opening of the oven door, machine noise, part weight, and the use of long tongs for part handling.
- The dies used in the extrusion and upset operations must be removed to repair wear every 2000 operations; the block and final forge dies must be corrected for wear every 1000 cycles.

PART 3: CDM IMPROVEMENT PROCESS

The CDM improvement process used by W-E for manufacturing systems and work cells has three steps: assessment, simplification, and implementation with performance measures. Before starting any improvement process, W-E executives verify that the following conditions are present:

- Management consensus favors the project and demonstrates full support.
- Employees who could be displaced as a result of the project understand the company's plans for them in the future.
- Management and union issues are fully resolved.
- All levels of management in the area targeted for the improvement process are represented in the change process.

In the CDM plan, the *assessment* was performed at the corporate and division levels by teams of employees representing all areas. The results of the assessment included (1) an analysis of the technology available and the technology currently used by W-E; (2) the educational levels and technical skills in all employee classifications; (3) an analysis of the manufacturing processes and strategies used by W-E and by their competitors; and (4) a comparison of the order-qualifying and order-winning criteria of the W-E products with other brands.

In general, *simplification* in CDM focuses on the *elimination of waste;* however, the simplification procedure is dictated by the production process under study. Every process has a number of variables that determine the success of the final product. The variables have a target value that the process holds within some tolerance; however, waste results when the process operates outside that tolerance. For example, if a machined part falls below the minimum allowed variation in the cut dimension, the part is scrapped. Waste also results when the values for process variables are not chosen carefully. For example, the drying time of a sealant has a specified value, but the time will vary, depending on ambient conditions. The drying time is considered as waste, as no value is added to the product while the sealant dries. Waste could be eliminated in the sealant application by using a sealant with a faster drying time or by applying heat to reduce the drying time. The CDM project emphasizes the elimination of non-value-added operations in a process; in addition, CDM places equal emphasis on process variable *elasticity*. During the improvement process, all process variables must be checked for elasticity, that is, the ability to be changed to reduce waste while not changing the final product. The elasticity of a process variable is often determined by the manufacturing process. In the sealant example, the sealer could have a much faster drying time if it were applied more swiftly by a robot rather than a human operator.

Implementation with performance measures, the last part of a CDM improvement process, is project specific. The performance measures used to determine the level of project success come from the order-winning goals for the product. Based on information gathered in assessment, simplification, and implementation, a five-step design process is initiated to determine the production and automation hardware and software, machine and process modifications, facility changes, and training required.

Five-Step Design Process

To achieve the goals of CDM, changes in the design, production control, and manufacturing areas are necessary. As a result of these changes, some new production and automation equipment is usually needed, and a design process to successfully implement the hardware and software is necessary. W-E uses the following process:

1. Set up the project team. The project team typically should include representation from the following areas: *product design, quality assurance/control, manufacturing process design, production standards, economic justification, plant engineering, manufacturing, maintenance,* and *human resources.* Typically, the team is led by a representative from manufacturing engineering and could include outside consultants whose expertise is not available in the organization or team.

2. Define the improvement project objectives. The objectives support the product order-winning and order-qualifying criteria by addressing *quality, product cost, productivity, safety (product and employee), production flexibility,* and *production agility.* Product cost includes inventory and setup improvements, and production agility addresses the ability to move quickly into new market areas.

3. Select the appropriate process for improvement. The process selected for improvement should provide the necessary return on investment and have a high likelihood for successful implementation. In addition, the process targeted should be improved in one or both of the following areas:
   - Order-qualifying or order-winning criteria
   - Customer or employee safety

4. Design the work cell or system. The design process focuses on the following issues:
   - A study of the production process from three perspectives: *technical, economic,* and *human*
   - Selection of the most appropriate automation hardware and software
   - Integration of the hardware in the work cell or system
   - Integration of the work cell or system information with other enterprise systems

5. Build and operate the work cell or system. The tracking of work cell or system performance against design goals is a part of this final step.

PART 4: THE AUTOMATION PLAN

After the initial meeting, Bill met with Marci, who was also a manufacturing engineer, to get another opinion on the current problems in the blade production area and to indicate that he wanted her to take a major role in the automation project. The team needed some help from the quality area, and her knowledge and experience in that area impressed Bill. It was immediately apparent that her recent experience in the MRP II system startup was going to be a benefit to this project. She agreed to brief the entire team on some of the critical issues at the next meeting.

The W-E project teams usually have a representative from finance to assist in the justification. Bill wondered why Roger didn't include someone to cover that area. Tom O'Brien, a golfing friend in cost accounting, might lend a hand in that area if necessary. Other than that, all the other areas were covered. It was getting late, and Bill needed to work through the design process, take a casual stroll through the blade production area, and make some to-do lists for the team and the meeting. He better let Sandy know he would not be home for dinner.

Early the next morning Bill met with Ellen in her office because she could not attend the 9 meeting. He was glad to hear that all the production workers would be briefed on the plans for automation, starting with the first shift. Previous work on continuous improvement teams indicated to Ellen that the operators were willing to help in the process as long as the implementation would not produce layoffs. Most of the forge work-cell operators were male because of the strength required to hold the parts with the long tongs. Because of the stress and working conditions in these cells, the union was willing to agree to automation. The coating operators were primarily female, and many liked the work, so there may be a problem in this area for future automation.

Everyone arrived early for the meeting. After briefing the group on his earlier meeting with Ellen, Bill said, "Judging from my e-mail yesterday, you have some questions about this project. Before we open up the discussion, I would like Marci to give us an overview of the production issues in the blade line."

Marci stood up, pushed copies of a report on blade production to the middle of the table, and paused. When everyone had a copy of the report, she glanced at her notes and said:

> I did this report about 12 months ago when we were looking into production control software for mid-term and near-term scheduling of production on the blade production line. What we found out was that the investment in software could not be justified until a number of production issues present on the line were addressed. You can read the details later, but now I will just cover the major points. The major problems were quality and productivity. Let me discuss the quality issue first.
>
> The present system has a quality check after extrusion and upset forging using a sampling technique; then the cross section of each airfoil is measured after the deburring operation. The variation of airfoil cross sections in the final test is too wide. We found that the problem results from the following: die wear, nonuniform metal flow during forging, dimensional variation in the part during the upset operation, and, to some extent, raw material variation and slug preparation. The next question was what caused these problems. Everyone working in this area knows that getting consistently good forged parts is as much an art as it is a science; however, we were able to verify several process variables that created the problems. We found that die wear increased significantly when the cycle time between cold part and forged part varied moderately. The thickness or inconsistent coverage of the lubrication coat on the block forged part had a measurable effect on final forge results.
>
> The second issue was productivity. We are under pressure from our turbine and jet engine divisions to pass along a lower cost to them for the blades we ship. They can get blades outside the company at more competitive prices. In addition, the divisions want more frequent shipments of smaller lot sizes on all of the parts. A major problem here is the 15 to 20 percent downtime per shift due to die change, preheating dies, maintenance, and production problems. Remember, this production process has not changed significantly in the past 20 years.
>
> Some quality improvements have been made by the continuous improvement teams, but we can't control the cycle time variation that occurs from the

start of the shift to the end. In addition, quality from shift to shift still varies more than we can accept. Work on the downtime problems has reduced it to 15 percent, but that is still too high. We need lower and more consistent cycle times at every cell. Are there any questions?

For the next 45 minutes, the team engaged in an active discussion of the production issues and the challenge of automating the line. After all the major concerns and questions were voiced, Bill brought the team back to his next agenda item, the automation design process. He suggested that everyone locate the five-step process in their handout from the last meeting. He waited as folders opened and pages turned and then said:

As you can see, the process provides guidance but gives us the flexibility we need. Our team has representation from most of the areas listed in step 1 except quality assurance and economic justification. Marci has agreed to represent the quality issues, and I think I can get the help we need to develop the business plan and justification. Based on Marci's report and on discussions I had with Roger, our project objectives are improved *quality* and *productivity*. The productivity improvement should address lower product cost and smaller production lot sizes. As always, consideration for customer satisfaction and employee safety are a top priority.

The next major decision is where to start in the blade production line. The process we select must provide the necessary return on investment, and I see no way to get the 2-year payback without eliminating direct labor cost from the process. Therefore, I suggest we consider robotics as part of the automation solution. In addition, Roger is counting on a highly successful implementation. So we will use a survey to rate the automation opportunities in all the production areas in the blade line.

Bill passed copies of the survey (see Table 3–1 on p. 112) to both sides of the table and gave everyone time to review the survey. Judging that everyone was ready to go on, Bill continued.

Marci and Mike, I would like you to help me do the surveys of the work cells as a first step in this process. I suggest we each take a different shift since we may get more information by seeing as many operators as possible. After all production cells are surveyed, those that received six *yes* answers will be ranked by the team based on project objectives, best return on investment, and likelihood of a successful project. If robots are not the most effective path, then we should know that early in the project.

After we have selected the work cell or cells for the project, the design of the cell or system will start. We have a lot of work before we reach that point. While we are doing the survey, Jerry, I would like you to put together a report on the history of maintenance problems on all the major production machines used in the blade line. Tom, we will need drawings showing the layout of all the production work cells in the line. As you do these, keep in mind that three-dimensional models of the automated cell are necessary later in the design process.

Table 3–1 Plant Survey.

Robot applications—initial plant survey

Answer the following questions for each workstation in the plant survey:

1. Can inspection by operators be eliminated from this workstation?   Yes   No
   It is difficult and expensive to include parts inspection in a robot work cell.

2. Is the shortest machine cycle 3 seconds or longer?   Yes   No
   Robot speed is limited. Human operators can work faster than robots when demanded by the process.

3. Can the robot displace one or two people for three shifts?   Yes   No
   If the average robot project costs $100,000, then it will be necessary to save the cost of one or two operators for three shifts to get a 1- or 2-year payback.

4. Can the parts be delivered in an oriented manner?   Yes   No
   Picking parts from a tote bin is easy for humans but very difficult for robots. If the parts can come oriented for easy robot pickup, then robot automation is possible.

5. Can a maximum of 6 degrees of freedom do the job?   Yes   No
   Robots have one arm that moves through a restricted work space compared with the two-armed human. A single-armed robot must be able to do the job.

6. Can a standard gripper be used or modified to lift the part or parts?   Yes   No
   The tooling is a major part of the work-cell expense. The simpler the tooling can be, the greater the likelihood of a successful project. Also, the weight of the part plus gripper must be consistent with the robot's capability.

Bill paused to let everyone finish writing, then said, "How long will it take each of you to get that material ready?"

After thinking for a few seconds, Jerry said, "I think a week will be necessary because some of the information is scattered across several departments.

"Some of the work cells are already in the CAD system, so a week for the rest is about right for me, also," added Tom.

"I have no other agenda items. Any other questions?" Bill looked at both sides of the table and then said, "We will meet again in one week . . . I will let you know the time and place." ●

3-9 SUMMARY

The use of stand-alone automated work cells did not deliver the productivity results needed by many enterprises to remain competitive. The poor results were traced to three sources: isolation of the automation from the production system, deficiencies in the work-cell design process, and insufficient training and preparation of people. A CIM implementation process that includes assessment, simplification, and implementation with performance measures will overcome most of the deficiencies with the earlier automation implementation.

The assessment step includes detailed self-study of the enterprise in three areas: technology, human resources, and process systems. In each case, the assessment documents what the enterprise does well and what needs improvement. The second step, simplification, focuses on the elimination of waste. Waste is every operation, move, or process that does not *add value* to the final product. The first two steps prepare the enterprise for the CIM implementation; step 3 builds the system. Implementation with *performance measures* implies that the work cells and integrated systems are purchased and installed, but measurements of system performance must also be included. The six key measurement parameters are product cycle time, inventory, setup times, quality, employee output/productivity, and continuous improvement. The specific process used by companies as they address the implementation of CIM principles will vary depending on the conditions present and the corporate culture. However, some of the elements are support from the top, multifunctional teams, willingness to study all processes, valuing every employee's input, and accepting a 3- to 5-year payback.

Flexibility in manufacturing is often described as (1) the ability to adapt to engineering changes in the part; (2) the increase in the number of similar parts produced on the system; (3) the ability to accommodate routing changes that allow a part to be produced on different types of machines; and (4) the ability to rapidly change the setup on the system from one type of production to another. Flexible manufacturing takes two forms: flexible manufacturing systems and flexible manufacturing cells. An FMS is defined as one manufacturing machine or multiple machines integrated by an automated material handling system, whose operation is managed by a computerized control system. An FMC is defined as a group of related machines that perform a particular process or step in a larger manufacturing process.

Fixed automation systems and machines have little flexibility for changes in manufacturing process or products. However, they do have the ability to manufacture large volumes of discrete parts to economically meet production demand for some products. Fixed automation systems are implemented in two configurations: in-line and rotary. The in-line fixed automation systems include a linear work flow through a series of workstations. Rotary-type fixed automation locates the production stations around a circular table or dial. As a result, this type of production system is called an indexing machine or a dial index machine. A rotary system is limited to smaller workpieces and fewer production stations than the in-line type.

The implementation of automated work cells forces two decisions early in the design process: (1) the use of current production hardware versus the purchase of new process machines, and (2) the adoption of fixed versus flexible automation. In the first instance, a choice must be made between adding automation to existing production machines or building a completely new production system. The second decision requires a choice between flexible and fixed automation.

QUESTIONS

1. Why were the islands of automation developed in the 1970s unable to perform to expectations?
2. How does the CIM development process correct the deficiencies present in the islands of automation?

3. What is assessment, and what areas of the enterprise are affected?
4. What is the focus of the educational component of assessment?
5. How is an enterprise changed by the simplification process?
6. What does implementation with performance measures imply?
7. Describe the six key measurement parameters used in CIM implementations.
8. Why do some companies avoid the CIM acronym when they start an improvement process?
9. What is usually considered the most critical element for the success of a CIM implementation?
10. Name the types of manufacturing systems in which robots are most likely to be used.
11. What does flexibility in manufacturing mean?
12. Compare and contrast FMCs and FMSs.
13. How are the automated systems installed at Technophone and TRT Thomson similar? How are they different?
14. What are the major differences between flexible and fixed automation systems?
15. Compare and contrast in-line and rotary-type fixed automation systems.
16. How do companies justify the use of existing production machines in an automation project?
17. What criteria are used in the selection of fixed automation over a flexible automation cell?

PROBLEMS

1. Using the cycle times for the extrusion, upset forging, block forging, and final forging cells, calculate the weekly production per cell for a three-shift operation assuming a 20 percent downtime per shift.
2. How much did the weekly production increase when the uptime was improved to 85 percent?

CASE PROJECTS

1. Using the case study description of the West-Electric blade production line and any other resources available, complete an initial plant robot survey for all the production cells.
2. Based on the results of the robot survey of the work cells and general case data, rank the work cells as follows:
   a. Least difficult to most difficult to automate
   b. Work cell you would recommend for automation first; the second cell area you would automate

3. How did the information and general case data provided by Roger Walker and Marci Hatcher influence your decisions in question 2b?

4. What step(s) in the blade line will be the most difficult to automate?

5. Was Bill Baxter correct to assume that the project had to include robotic cells?

6. What are the major reasons that the cycle times in the forging cells are not uniform? Could the problem be eliminated without automation?

7. What advantages would robots provide if they are used in the cells?

8. How can the team integrate the production on the entire line and avoid the problem of islands of automation?

9. Where is W-E in the three-step CIM process? What are the next steps for the W-E team after the selection of the work cell(s) for automation?

End-of-Arm Tooling

4-1 INTRODUCTION

Studies indicate that the process of joining two mechanical parts in an assembly operation uses an operator's sense of touch to a much greater degree than the sense of vision. This dominance of the tactile underscores the importance of the human hands in all phases of manufacturing and assembly of production goods. It also emphasizes the demand placed on the *gripper* or *end-of-arm tooling* of an industrial robot if it is to perform many of the production duties of its human counterpart. Duplication of the human hand with its ability to grasp, sense, and manipulate objects remains one of the most difficult tasks facing the designer of end-of-arm tooling. Automation researchers are working on a robot hand with three fingers that can grasp irregularly shaped objects, but widespread application of this type of tooling is years away. This chapter investigates the types of robot tooling currently in use and covers the terminology involved in end-of-arm, or *end-effector*, tooling.

In general, an end effector is the tooling or gripper that is mounted on the robot tool plate. The function of the gripper is to hold the part as the robot presents it to the tool for work to be done, or to hold the tool as the robot moves it to work on the part. For example, a part held in a gripper could be positioned under a numerical control (NC) drill, thereby producing a hole; or the end-of-arm tooling could hold a drilling mechanism, with the robot producing the holes in the same manner as a human operator using a hand drill.

The end-of-arm tooling used in a robot work cell should exemplify all five of the following characteristics:

1. The tooling must be capable of gripping, lifting, and releasing the part or family of parts required by the manufacturing process.
2. The tooling must sense the presence of a part in the gripper, using sensors located either on the tooling or at a fixed position in the work cell.
3. Tooling weight must be kept to a minimum since it is added to part weight to determine maximum payload.
4. Containment of the part in the gripper must be assured under conditions of maximum velocity at the tool plate and loss of gripper power.
5. The simplest gripper that meets the first four criteria should be the one implemented.

As a result, one of the major design problems associated with robotic work cells is the selection and design of the end-of-arm tooling.

4-2 CLASSIFICATION

The end-of-arm tooling used on current robots can be classified in the following three ways: (1) according to the method used to hold the part in the gripper; (2) by the special-purpose process tools incorporated in the final gripper design; or (3) by the multiple-function capability of the gripper. The first category of gripping mechanisms includes standard mechanical pressure grippers, tooling using vacuum for holding or lifting, and magnetic devices. The second classification of tooling includes drills, welding guns and torches, paint sprayers, and grinders. The third type of gripper tooling includes special-purpose grippers and compliance devices currently in use.

Standard Grippers

Standard grippers can have two different closing motions, *angular* or *parallel*, and can have *pneumatic, hydraulic, electric,* or *spring* power for closing and opening. The action of the angular and parallel devices is illustrated in Figure 4–1; two standard off-the-shelf grippers are pictured in Figures 4–2 and 4–3. In most applications the gripper base is purchased with an actuator mechanism and jaw mounts for the fingers. The gripper in Figure 4–2 has each special-purpose finger attached to a jaw mount with two screws. The geometry necessary for the fingers was determined by the cell designer, and the finger parts were machined for the application. Some grippers are supplied with blank jaws that can be removed and machined into the configuration required for the application. For example, the gripper on the left in Figure 4–3 has blank jaws, and the others have the blanks machined for specific applications. Starting with standard off-the-shelf grippers and modifying the jaws for the application is the most cost-effective approach. A standard gripper

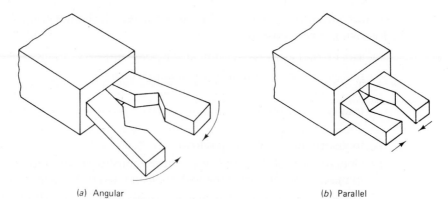

(a) Angular   (b) Parallel

Figure 4–1   Standard Angular and Parallel Grippers.

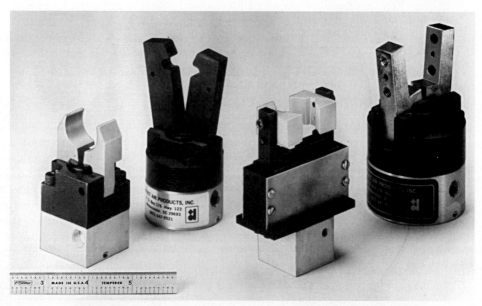

Figure 4–2   Angular Gripper Air Close Spring Open.

Figure 4–3   Typical Modifications to Blank Jaws. (Courtesy of Compact Air Products, Inc.)

costs between 4 and 8 percent of the cost of the robot; small blank jaw pneumatic devices are priced under $1000. If a specially designed end-effector is required, then the cost for design and fabrication often exceeds 20 percent of the total costs for the robot system.

The gripper must be closed and opened by program commands as the robot moves through the production operation. The robot controller supplies the electrical signals that result in the gripper's action. Most grippers are opened and closed with a pneumatic actuator (Figure 4–3); however, in limited applications hydraulic or spring power is used. In some cases, grippers are spring opened and power closed, as illustrated in Figure 4–2; less frequently, grippers are spring closed and power opened. Each of the types has advantages and disadvantages based on the specific application. The power-closed and spring-opened type, for example, only require power to close the gripper when a part is lifted and moved, which offers the advantages of simpler control and less power consumption. However, a part will be dropped or thrown if gripper power is lost while the robot is in the process of moving the part.

Grippers use both exterior features and interior geometry of the part to pick it up in an application. The two grippers in Figure 4–4 illustrate how the blank jaws of a standard gripper are machined to support external and internal part gripping. At the left in Figure 4–4, the jaw blanks were cut to grip the shaft around the external surface; however, in Figure 4–3 (second from right) the jaws were designed to pick up the shaft from the end. The variation in gripper jaw or finger design is as different as the types of parts handled.

Figure 4–4   Interior and Exterior Gripping. (Courtesy of Mack Corporation)

Figure 4–5 Three-Finger Gripper. (Courtesy of Compact Air Products, Inc.)

Another variation in standard grippers is the number of jaws or fingers used to grasp the part. Most applications use two-finger grippers. Three-finger grippers (Figure 4–5) and four-finger grippers (Figure 4–6) are used when parts need to be centered by the gripping process. A three-fingered gripper, such as the one in Figure 4–5, would be fitted with fingers and would be used to handle large, flat circular parts by grasping the edge in three places. The four-finger grippers in Figure 4–6 center the part in both horizontal and vertical directions when the jaws close on the part.

Electrically powered jaws are also used in limited applications with a solenoid or dc servomotor providing the opening and closing action. Servo drives will be used more frequently when grippers are developed that can vary the pressure

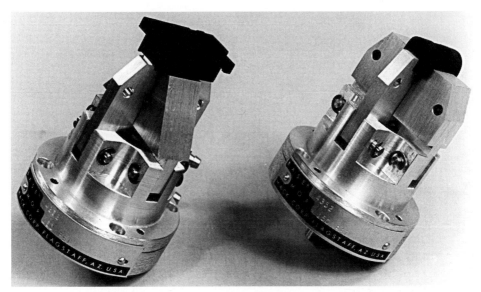

Figure 4–6   Four-Finger Gripper. (Courtesy of Mack Corporation)

applied to the object as it is grasped. This would permit the same gripper to lift either a steel ball or an egg without dropping the ball or breaking the egg. Tactile or touch sensing must be developed beyond its present state, however, before this type of gripper will be practical for the factory.

Vacuum Grippers

Vacuum is used as the gripping force in many tooling applications. The part or product is lifted by *vacuum cups*, by a *vacuum surface*, or by a *vacuum sucker gun* incorporated into the end-of-arm tooling. The lifting power is a function of the degree of vacuum achieved and the size of the area on the part where the vacuum is applied.

Most frequently, vacuum grippers use suction or vacuum cups to hold the desired part. The gripper can have a single vacuum cup, as illustrated in Figure 4–7, with a gripper picking up sheet metal plates from a stack. A proximity sensor is included to tell the robot when it has reached the top of the stack so that the robot can stop, apply the vacuum to the suction cup, and pick up the sheet. This system is capable of moving a variety of plate sizes in stacks of varying heights. Figure 4–8 shows a gripper with a multiple pattern of pickup cups that could lift large sheet metal plates, plastic sheets, plywood panels, or even large cardboard boxes. If each vacuum cup or selected groups of cups are individually controlled, then the gripper can lift variously sized sheets by activating only the cups required for the sheet size being lifted. It is this type of system flexibility inherent in a robot cell that makes it ideal for the automated factory.

The lifting capacity of the vacuum type of gripper is directly related to the pressure of the air surrounding the vacuum cups. Figure 4–9 provides additional

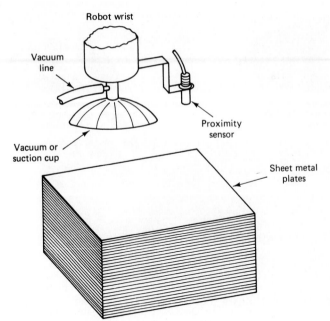

Figure 4–7 Vacuum Cup System to Unstack Sheet Metal Plates.

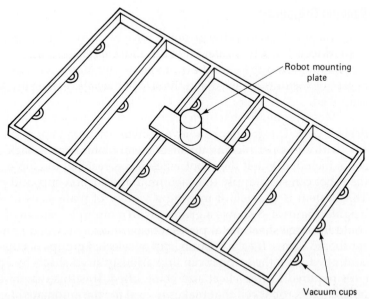

Figure 4–8 Multiple Vacuum Cup System to Handle Large Sheets of Material.

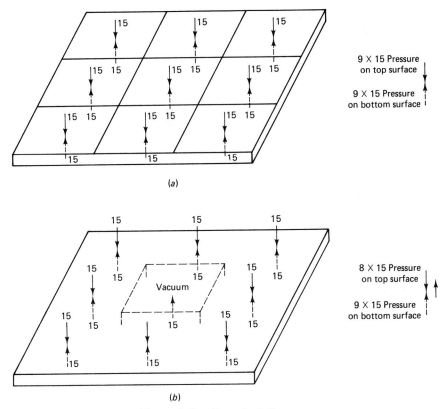

Figure 4–9 Atmospheric Pressure.

background on the principles involved in this lifting technique. The *atmospheric pressure* at sea level on the surface of an object is 14.7 pounds per square inch, but for simplicity in this example 15 pounds is used. Assuming that the thin sheet-metal plate in Figure 4–9a is under normal atmospheric pressure, the force on each side of the metal is 135 pounds (9 square inches × 15 pounds of pressure). Now if the vacuum cup in Figure 4–9b occupies 1 square inch of surface on the top side, then the effective force on the top surface is only 8 × 15, or 120 pounds. The force on the bottom surface is 15 pounds greater, and the sheet metal will be held against the vacuum cup by the difference in the forces. For example, a 3-inch diameter suction cup has 7.07 square inches of surface area and could, under perfect conditions, lift approximately 15 pounds for every square inch, or 106 pounds total. However, a nearly perfect vacuum is not easy or economical to achieve; therefore, a partial vacuum is used, and the size of the vacuum cup is increased. The lifting capacity of the gripper is unchanged, but the cost is significantly reduced. For example, a pneumatically powered vacuum generator and vacuum cup, available for less than $150, has sufficient lifting capacity for small and medium-sized parts. These pneumatically powered units use the venturi principle to create a partial vacuum. The surface of the part determines the quality of seal between the vacuum cup and the

Figure 4–10   Vacuum Gripper.
(Courtesy of GCA Corporation)

part; as a result, the surface of the material being lifted has the greatest effect on the total weight that can be raised. The smoother the finish on the surface, the closer the actual lifting weight is to the maximum. Figure 4–10 illustrates one type of vacuum cup available, and Figure 2–17 on page 48 shows a system of vacuum grippers used to install rear window glass during automotive assembly.

Vacuum Surfaces

Vacuum surfaces are just an extension of the vacuum cup principle. In some material-handling applications the product to be lifted is not ridged enough for vacuum cups to be effective. For example, robots are used to make composite material by building up multiple layers of graphite fiber cloth and resin. To lift the cloth into place a vacuum surface such as the one illustrated in Figure 4–11 is used. Note the flat surface that forms one side of a vacuum chamber. Each hole in the vacuum surface provides a small lifting force so that the flexible cloth would be held against the vacuum surface from many points as the graphite fiber cloth is moved into place. Any flexible material not too porous can be lifted effectively in this manner.

Vacuum Suckers

The use of *sucker guns* in the production of nylon fiber is well established. The tool consists of a wand that is capable of sucking up a nylon end or a thread line as it is produced, thus permitting the operator to thread the machine. This type of end-of-arm tooling could be very useful for applications in which robots must handle material in the form of thread, line, or fine wire.

Magnetic Grippers

Parts that contain ferromagnetic material can be lifted with an electromagnet mounted on the robot tool plate. Figure 4–12 is a picture of a *magnetic gripper.*

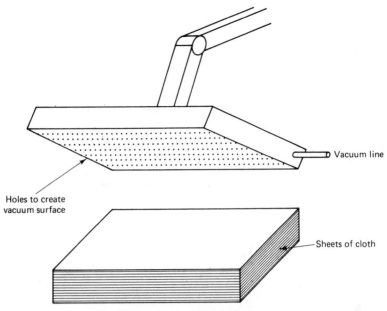

Holes to create
vacuum surface

Vacuum line

Sheets of cloth

Figure 4–11 Vacuum Surface.

Figure 4–12 Magnetic Gripper.
(Courtesy of GCA Corporation)

Air-Pressure Grippers

Fingers, mandrel grippers, and *pin grippers* form a group that uses air pressure to grip parts. The fingers, also called *pneumatic fingers,* have a hollow rubberlike body with a smooth surface on one side and a ribbed surface on the opposite side. With pressure applied to the inside of the hollow body, the finger deflects in the direction of the smooth side. Figure 4–13 illustrates this process,

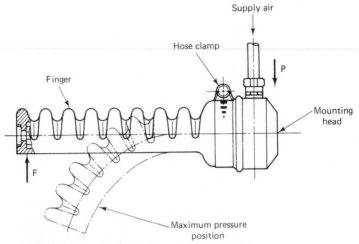

Figure 4–13 Pneumatic Finger.

Figure 4–14 Finger Gripper. (Courtesy of GCA Corporation)

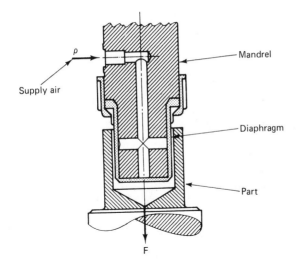

Figure 4–15 Mandrel Gripper.

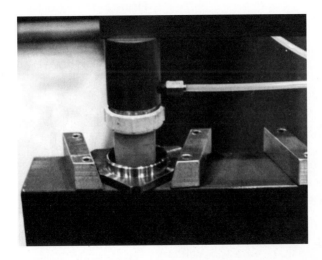

Figure 4–16 Mandrel Gripper. (Courtesy of GCA Corporation)

and Figure 4–14 shows a pair of parts being lifted by robot grippers with two pneumatic fingers.

Mandrel grippers are inside grippers with an airtight flexible diaphragm mounted to a mandrel. Figure 4–15 shows a section view of a circular mandrel gripper inside a part. When air under low pressure (25 to 35 psi) is forced into the port, the diaphragm expands and traps the part. Figure 4–16 shows a mandrel gripper (diaphragm is light gray) on robot end-of-arm tooling moving into a part.

Pin grippers, Figure 4–17, are similar to mandrel grippers except that the part is gripped from the outside. The gripper fits down over a round pin, then the diaphragm is expanded inside the ridged housing, and the part is trapped. The primary advantage offered by air pressure grippers is the gentle variable force

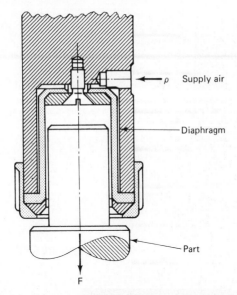

Figure 4–17 Pin Gripper.

Figure 4–18 Bellows-Type Parallel Gripper.

applied to the part being held. Figure 4–18 illustrates an air-pressure parallel gripper from Festo designed to lift blocks of raw material stock. Note the expansion bellows on the inside of the finger.

Special-Purpose Grippers

The expanding robot application base supports the development of off-the-shelf grippers to fill special applications. As a result, the largest classification of grippers now in use falls into the category of special-purpose tooling. In addition

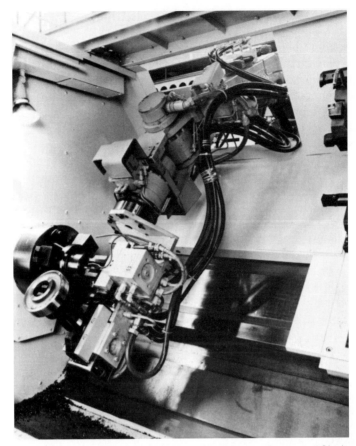

Figure 4–19 Robot Loading a Cinturn Turning Center. (Courtesy of Cincinnati Milacron Corp.)

to those available off the shelf, many more are fabricated by the robot end user. Often the user starts with a basic pneumatic or electric mechanism available in the industry and modifies the gripper to do the special job. In almost every case special gripper fingers must be fabricated to hold parts for specific jobs. Figure 4–19 shows two internal grippers mounted at 90 degrees so the robot can unload a finished part from the chuck and load raw material in one trip to the turning center.

Again, robot flexibility is responsible for the large variety of gripper designs. It is often a simple matter to identify a secondary task that the gripper can perform during the production cycle in addition to the primary parts-handling or assembly responsibility. Examples of secondary tasks that can be incorporated into a standard gripper include hooks for lifting or turning parts over, air guns for cleaning, and sprayers for lubrication of parts and dies.

4-3 GRIPPER SELECTION AND SYSTEM INTELLIGENCE

In setting up any manufacturing system, some criteria must be used in the selection of equipment. When robots are included in the automation, an analysis of each production sequence will establish the requirements of the gripper for each step of the manufacturing process. Every sequence in the manufacturing process should be examined and the relative difficulty established. The robot system capability, including the end-of-arm tooling, must be equal to or greater than the most demanding sequence in the process. In this way, the robot system intelligence level will always match the need of the production requirement.

There is another advantage to analyzing the requirements of each sequence in the manufacturing process. Often only one sequence is classified as difficult and that sequence dictates a robot system with far greater capability than is required for most operations. As a result of such an analysis, the demanding sequence can be identified and possibly modified to permit a lower cost implementation.

4-4 SPECIAL-PURPOSE TOOLS

Grippers are also designed to hold power tools in the same way that human hands do. Those most frequently used include drills, welding guns, glue and sealer dispensers, spray guns, grinders, and sand blasters. Robots are often used for jobs that

Figure 4–20   AdeptOne Robot at Chocolat Frey of Buchs, Switzerland. (Courtesy of Adept Technology, Inc.)

fulfill the three Ds: dirty, dangerous, and dull. A human operator using any of these tools must wear equipment to protect his or her eyes, eardrums, or respiratory system from harm and often requires counterbalance devices to overcome the weight of heavy tools. As a result, these jobs are ideally suited for robots.

4-5 ROBOT ASSEMBLY

Assembly applications for robots will grow at a rate faster than all other current or future application areas simply because assembly requirements occur in every segment of manufacturing. As gripper dexterity and part manipulation capability increase with developing technology, and as microprocessor systems are designed to control more complex grippers, the number of assembly applications will increase. Figure 4–20 shows an AdeptOne robot inserting chocolates into liners; note that the candy is delivered on a conveyer in random locations. A vision sensor interfaced to the robot is used to locate the chocolates on the conveyer.

Grippers designed to be used in assembly operations have special needs that must be satisfied. For example, parts must be moved into place for the assembly operation, fasteners must be put into place, and assembly tools must be applied. In all of these operations the compliance between the parts must be considered.

4-6 COMPLIANCE

Compliance deals with the relationship between mating parts in an assembly operation. For example, if a pin must be fitted into a hole, alignment between the hole and the pin must be achieved. Because it is impossible to guarantee perfect alignment between the robot gripper that holds the pin and the fixture holding the part, compliance techniques must solve the slight misalignment problems between mating parts during assembly. Compliance is defined as follows:

> *Compliance means initiated or allowed part movement for the purpose of alignment between mating parts.*

When a pin is inserted into a hole, three types of contact can occur during the insertion process: *chamfer, sliding,* and *two point.* The chamfer contact occurs when the pin is not perfectly aligned with the hole and hits the chamfered edge of the hole (Figure 4–21a). If the assembly system is not perfectly rigid, the pin will rotate slightly and start to slide along one side into the hole (Figure 4–21b). If the misalignment is severe, the pin will make a two-point contact, with the base of the pin hitting the far wall of the hole (Figure 4–21b). Two-point contact on insertion often results in damage to both the parts and the assembly system. Problems with mating part alignment in assembly and other applications are resolved using *active* and *passive* compliance techniques.

The initiated or allowed movement in each technique will be lateral, rotational, or axial about a center of rotation. Figure 4–21 shows the three basic types of movements and the misalignment that must be eliminated by each. In Figure 4–21a, lateral part movement is required for mating. A rotational correction is

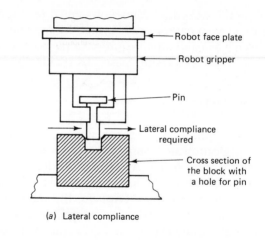

(a) Lateral compliance

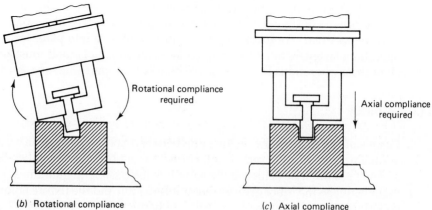

(b) Rotational compliance

(c) Axial compliance

Figure 4–21 Compliance.

necessary for proper part insertion in Figure 4–21b. Devices that provide lateral and rotational compliance are commercially available. The axial compliance illustrated in Figure 4–21c is not necessary for proper alignment but is added when the pin must be flush with the bottom of the hole. Using axial compliance, the robot could move in the axial direction, slightly past the point where the pin is seated. In these figures, each compliance condition was considered separately, but in operation they all may be present at one time, and the compliance technique employed must continue to function. Without compliance the entire system is rigid, so that force applied by the robot when two-point contact is present will result in damage to the two mating parts.

Active Compliance

Active compliance systems measure the *force* and *torque* present when the robot performs the programmed task; as a result, they are often called *F/T sensing* systems. Force sensing allows the robot to detect changes and variations in the work-

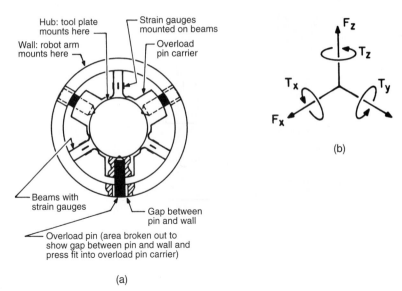

Figure 4–22 Force/Torque Transducer (a) and Vectors (b). (Courtesy of Assurance Technologies, Inc.)

piece or tooling and adapt the program to correct them. In F/T sensing systems, the robot moves are based on a given force instead of a programmed point.

F/T sensing uses an adaptor placed between the gripper and the robot tool plate to measure the force and torque caused by contact between mating parts. The adapter has an outside ring or wall (Figure 4–22a) that attaches to the robot tool plate. A center hub is connected to the ring by three webs or beams (Figure 4–22a) that have strain sensors attached. With the gripper attached to the hub, any force or torque on the gripper as a result of a programmed action causes the webs connecting the hub to the ring to deflect. The deflections are measured by the strain sensors and converted into three force vectors and three torque vectors (Figure 4–22b) that are used by the robot to adjust the programmed motion. Note that overload pins attached to the hub prevent damage to the sensor if too great a deflection is experienced.

The gear assembly application, pictured in Figure 4–23, shows the F/T sensor active compliance device between the gripper and tool plate of a robot. Adept Technology, Inc. has integrated the control of an F/T sensor into the robot controller; as a result, the robot can read and react to contact with the tooling in less than 3 milliseconds. In the Adept integrated system, the following force strategies are available: move until the specified force threshold is exceeded, move in a defined spiral search pattern to find a hole, move to verify assembly, and move to cause a snap-fit assembly. F/T sensors are available from 3/1 to 600/3600 (pound/inch-pound).

Passive Compliance

The second technique, *passive compliance,* provides compliance mechanically by means of a *remote center compliance* (RCC) device. The concept was originally

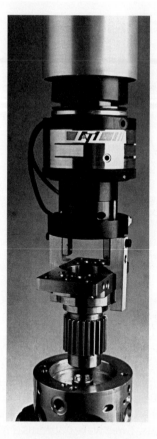

Figure 4–23   Force/Torque Sensor in Robotic Assembly (Courtesy of Assurance Technologies, Inc.)

developed in the 1930s to reduce vibrational stresses in radial aircraft engines by permitting deflections in the engine mounts that coincided with rotations about the engine's center of rotation. The theory of operation of RCC devices is a little abstract, but understanding the principle of operation is a minimum requirement for anyone using these devices in a work cell. The *center of compliance* (remember, compliance implies movement) is that point at which the entire compliance system including the gripper and part is considered to be concentrated and acting. The location of the center of compliance is determined by the design of the RCC device and depends on the location of the compliance elements and their orientation to the RCC device. As the name indicates, the center is usually remote from the device itself. Figure 4–24 illustrates this concept.

As we discuss the operation of the RCC device consult the design in Figure 4–24, which came from research work at the Charles Stark Draper Laboratory at MIT. The RCC device consists of three plates; the center plate is connected to the top plate with four rods and to the bottom plate with four additional rods. In operation, four rods, one on each corner, are used for lateral compliance (only two rods are shown in Figure 4–24), and four angled rods, one on each corner, are used for rotational compliance (again, only two rods are shown). The flexible rods allow the plates to move relative to each other and provide a combination of lateral and rota-

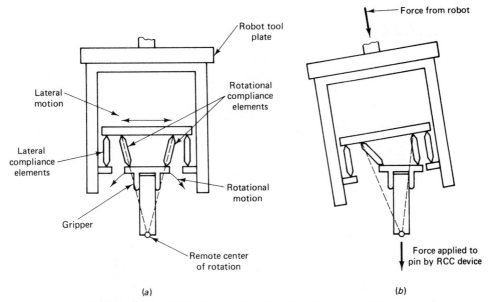

Figure 4–24 RCC with Lateral and Rotational Compliance Links.

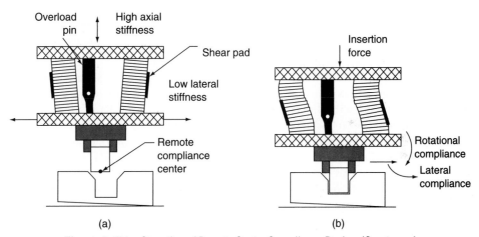

Figure 4–25 Operation of Remote Center Compliance Device. (Courtesy of Assurance Technologies, Inc.)

tional compliance; however, this device is rigid in the axial direction with no compliance provided.

Modern RCC devices, like the model illustrated in Figure 4–25 and pictured in Figure 4–26, consist of three to six shear pads sandwiched between two aluminum plates and shear pins for protection against excessive movement. This type of RCC design is called the *single-stage shear pad* device. The key to the performance is the construction of the shear pads of from 15 to 30 alternate layers of neoprene

Figure 4–26 Remote center Compliance Device.

or nitrile elastomer and metal shims. The upper plate is attached to the robot tool plate and the lower plate is attached to the gripper. The shear pads are stiff in the axial directions but highly compliant in the lateral and rotational directions. The slight angle of the shear pad causes the rotational compliant effect illustrated in the model in Figure 4–24. Note in the model and in the actual device (Figure 4–25) that the lower plate rotates because the increased angle on one shear pad reduces the distance between the upper and lower plates. Additional rotation is caused by the other shear pad, which moves to a more vertical position, thereby increasing the distance between the two plates. Study the action in Figure 4–25 until you understand the lateral and rotational compliance.

Two additional features, *shear pins* and *lockup,* are available on current RCC devices and improve the performance of the original design. The shear pins provide protection for the shear pads against the occasional and inevitable overloads that occur. The shear pin is attached to the top plant and is centered inside a hole in the bottom plate. If the lower plate tries to move an excessive distance, then the shear pin comes in contact with the lower plate and prevents further movement. The RCC device usually has three shear pins. The lockup feature removes the compliance from the RCC device whenever the device is not performing an assembly. The weight of the part and gripper could cause the RCC device to oscillate after a quick move to the assembly. The insertion operation would have to be delayed until the oscillation died out naturally. To overcome this problem, RCC devices have a lock-up feature that makes the device rigid for moves or for operations

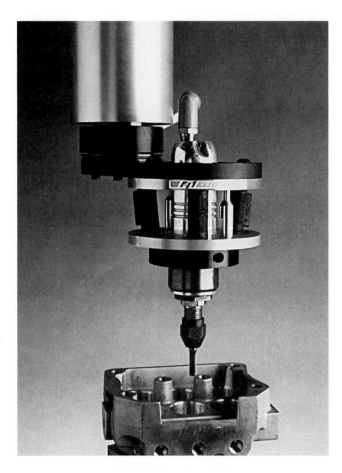

Figure 4–27 Remote center Compliance Device in a Deburring Application. (Courtesy of Assurance Technologies, Inc.)

where compliance is not desired. The lock-up capability reduces assembly cycle time and increases the operational life of the shear pads. Figure 4–27 shows an RCC device used in a pneumatic deburring application. The soft lateral compliance permits the burr to have a constant pressure against the part.

Selective Compliance Articulated Robot Arm. Another form of passive compliance is found in the SCARA configuration. The AdeptOne robot in Figure 2–18 uses SCARA technology to provide a variable tool movement for insertion compliance at programmed points. After the tool reaches the programmed point, the controller frees the servo system so the gripper can move freely over a selected distance in the X and Y directions. If the gripper attempts to move beyond the selected range, the servo system stiffens, and the gripper position is maintained.

In addition to the SCARA arms that initiated the selective compliance technique, some non-SCARA arms—for example, the Seiko RT-3000—have language commands that permit any of the axes to have compliance. The command for the Seiko is **FREE** <*axis*> where the argument is the letter for the axis where compliance is desired.

4-7 MULTIPLE END-EFFECTOR SYSTEMS

In most applications currently using robots, the machines have a single function to perform in the work cell, and only one end effector is necessary. As work cells become more complex and robots more sophisticated, however, the application of multiple gripper systems increases. The introduction of multiple grippers demands a standard interface at the robot wrist.

Wrist Interface

The wrist interface for flexible manufacturing must satisfy the following interface requirements:

- *Mechanical interface:* The robot must be able to change tooling under program control, and the integrity of the mechanical linkage must be as good as that experienced with threaded fasteners. The interface must provide both registration and orientation control from one tool to the next.
- *Electrical interface:* The electrical signals used for control of the tooling or the signals coming from sensors mounted on the gripper must be separated automatically when the tooling is changed. For example, the power to electrically powered tools must be disconnected and then reconnected as the tooling is changed.
- *Pneumatic interface:* The same rationale for a quick-breaking interface that was developed in the electrical area also applies here.
- *Replaceable or quick-change capability:* Future applications must provide for rapid tool change by the user and the robot.

The advantages of a standard wrist interface include these three: (1) the same tooling can be used on every robot in a manufacturing facility without special adapters for each robot; (2) no program editing or reprogramming is required when a worn or damaged gripper is replaced; and (3) off-the-shelf tooling for general applications can be developed and sold.

The Robotic Industries Association (RIA) is working on several standards that affect the mechanical interface area. One of those, ANSI/RIA R15.03, for industrial robots and robot systems—mechanical interfaces, has two components. The first is the circular mechanical flange interface standard and the second is the shaft mechanical interface standard. Development of standards will help move the robot industry toward interchangeable tooling.

Multiple-Gripper Systems

A multiple-gripper system is one that has a single robot arm but two or more grippers or end-of-arm tools which can be used interchangeably on the manufacturing process in the cell. These multiple tooling systems can be completely separate grippers or tools mounted to the fixture on the end of the robot arm. The tooling pictured in Figure 4–28 shows six pneumatic grippers with a combination of two, three, and four jaws to handle many tasks.

Figure 4–28 Multiple End Effector. (Courtesy of Mack Corporation)

In other applications, the robot tooling is designed to permit two or more grippers to be used on the same robot arm on an interchangeable basis. The discussion of the wrist interface indicated that interchangeable tooling would require an interface for mechanical, electrical, and pneumatic requirements. A quick change mechanism, developed by Assurance Technologies, Inc., satisfies the mechanical, electrical, and pneumatic requirements. The quick change device, shown in Figure 4–29, consists of a master plate that connects to the robot tool plate and a tool plate that attaches to the gripper. Study the figure and notice the parallel rows of fifteen electrical contacts and the eight pneumatic interfaces. The external connections for the electrical and pneumatic ports are visible as well. The interface sequence is illustrated in Figure 4–30. The interface uses pneumatic air only for joining and separating so that a loss of air pressure does not create a safety hazard. Device models have payloads that range from 5 to 150 kg, ten to seventy-four electrical connections, and from six to fourteen pneumatic ports. The current for the electrical interface is typically 3 amperes at 50 volts (higher capability is available on larger models). The pneumatic interface can operate from a vacuum up to 100 psi. The repeatability of the interface is ± 0.010 millimeters and would add to the repeatability present in the robot.

As work cells become more complex and costly, the concept of increased flexibility through multiple end-of-arm tooling makes good economic sense. Cost justification on this basis alone is difficult to establish in terms of person hours saved, but in many production situations the use of robot automation would not be possible if multiple grippers were not employed. For example, in a production situation in which the range of part sizes is great, the only way to avoid installing several robots is through multiple grippers. In an assembly application in which the number and variety of operations to be performed is

Master plate

Tool plate

Figure 4–29 Quick Change Device. (Courtesy of Assurance Technologies, Inc.)

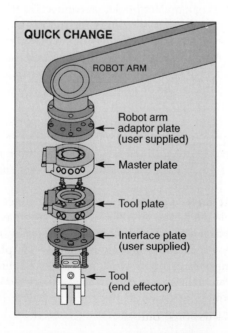

QUICK CHANGE

ROBOT ARM

Robot arm
adaptor plate
(user supplied)

Master plate

Tool plate

Interface plate
(user supplied)

Tool
(end effector)

Figure 4–30 Quick Change
Tooling System.

Figure 4–31 Multiple-Gripper System. (Courtesy of Fared Robot Systems)

numerous, it would not be economically feasible to have a robot and separate gripper for each operation.

The advantages of using multiple grippers include (1) increasing the production capability of the work cell; (2) reducing work-in-process time for the part, because it must be moved through fewer workstations; and (3) using the robot arm and controls more efficiently. In addition, many assembly applications would not be possible without the development of multigripper work cells. The primary disadvantage is the additional complexity that is added to the tool design problem. This also results in greater cost for the design and construction of the devices.

The multiple-gripper system pictured in Figure 4–31 is used in an assembly work cell. Each different tool would have a TCP offset value that would be changed in the robot program when the tool is changed.

CASE: CIM AUTOMATION AT WEST-ELECTRIC

As Bill drove to work he thought about the team meeting in the afternoon. Despite the still many unknowns, the project was falling into place. The initial work-cell surveys were complete, and he was more convinced than ever that robot automation was the best plan. There were few variations between Bill's survey of the first shift and the surveys of the second and third shifts done by Marci and Mike. From their initial discussions, Bill sensed that Marci wanted to be more aggressive in the initial integration of the cells, and Mike was taking a more conservative approach. Knowing Roger's interest in a successful implementation, Bill leaned toward the conservative approach; however, the team expressed consensus that robot automation was the correct direction. The report from Jerry on the condition of work-cell hardware supported the automation of existing production equipment. The forges were the oldest equipment in the cells, but they were in excellent condition. The ovens were only a few years old, and the internal parts shelf could be modified for automated operation. The automated coating operation was only 5 years old and was well maintained.

Marci and Mike felt the first project should be in either the extrusion or the upset areas, and Bill agreed. The only question was how much to integrate and to what degree to automate. This decision needed to be a team effort, so he gave everyone on the team a copy of Jerry's report, Ted's drawings of current work-cell layouts, and the initial work-cell surveys from all three shifts. He asked them to review the material and come to the meeting with first, second, and third choices for the initial automation project and rationales for their choices.

After briefing Roger at 9 on the status of the project, Bill started to review his handouts for the next step in the design process, the technical design checklist for the work cell and robot selection.

WORK-CELL TECHNICAL DESIGN CHECKLIST

A good design requires a detailed study of the cell and the factors that contribute to its success. The following checklist serves as a guide to identify the technical issues that must be addressed during the design of the cell. The checklist is divided into five major areas: *performance requirements, layout requirements, product characteristics, equipment modifications,* and *process modifications.*

Performance requirements

Cycle times	Tolerance of parts
Part handling specifications	Dwell time of tools
Feed rate of tools	Pressure on tools
Product mix	Maximum repair time
Equipment requirements	Malfunction routines
Human backup requirements	Allowable downtime
Future production requirements	

Layout requirements

Geometry of the facility	Service availability
Environmental considerations	Floor loading
Accessibility for maintenance	Safety for machines and people
Equipment relocation requirements	

Product characteristics

Part orientation requirements

Surface characteristics

Unique handling requirements

Gripper specifications

Size, weight, and shape

Inspection requirements

Equipment modifications

Requirements for unattended operation

Requirements for increased throughput

Maximum/minimum machine speed

Requirements for automatic operation

Process modifications

Lot-size changes

Process variable evaluation

Routing variations

Process data transfer

System integration

Data interfaces and networks

Hardware integration requirements

Data integration requirements

Interface requirements

Software integration requirements

ROBOT SELECTION CHECKLIST

A major step in the design process for a work cell that includes robotics is the selection of the robot system. After the technical issues for the robot work cell are resolved, a robot model is selected that matches application requirements. The following robot characteristics must be considered:

Positioning resolution, repeatability, accuracy

Arm geometry

Positioning flexibility

Maximum and payload dependent velocity

Compliance requirements

Force/torque sensing requirements

Cost

Vision integration

Work envelope size

Degrees of freedom

Maximum and rated payload

Downward force

Tool change requirements

Programming (on- and off-line)

Special option requirements

As everyone arrived for the meeting, Bill handed them a folder with the agenda attached and the checklist material enclosed. He let everyone exchange greetings, then started the meeting by saying:

The first thing we need to do is to define the first automation project. I will record each person's top three priorities on the flip chart, and we will see if there is a consensus on the cell to automate first. Please give your rationale for the choice, as well, so any information you used to reach your decision will be shared with everyone. I will ask the three of us who completed the survey to start off. Marci, would you go first?

Marci got out her notes and said, "My first priority is an integrated system that starts with slug coating and ends with production of the upset forged part. My second priority would be just the upset cell, and the third would be just the extrusion cell. My reason for starting with those two areas is that I got *yes* answers for every question on the survey for both of those cells," she said. "The inspection can be handled outside the automation, and the technical issues are not that tough. If we need to be more conservative and limit the automation to a single cell, then my choice on the individual cells would be the two I mentioned." She paused to look at some notes, and then said, "If we include the direct labor cost for nine operators in each product area and the increase in production, the project should have the cost justification numbers that Roger requires. Even if we keep one operator per shift to maintain stock and handle minor automation problems, we will still get the justification in under 2 years." Marci's look at Mike said she was finished and he had the floor.

Bill nodded, and Mike said, "I think we need to be conservative on the first attempt at automation, so I just have two priorities: first, the upset cell, and second, the extrusion area. We can economically justify replacing the operators with robots as long as we get a 20 percent increase in output." Mike looked at Marci and continued, "I'm not afraid of the automation or the challenge, but I would like to stay away from dealing with coated parts for the first project. The technical issues associated with the upset cell are not a problem with current technology in robots, sensors, grippers, and work-cell controllers." Mike looked at Bill to indicate that he was finished for now.

Bill decided to wait until the others listed their priorities before indicating his preference for the first project. Everyone agreed that the project should focus first on two blade operations; most picked the upset cell for the first attempt at automation. It was time for his priorities, and so he said, "Like many of you, I picked the upset as my number one choice because I think we can get the automation working more quickly and easily. In my briefing with Roger this morning, he mentioned again his interest in a more cautious first project, but one that built in the capability of broader integration in the future. Therefore, I would suggest that we start with the upset operation; however, I don't want to just replace operators with automation. I would like some suggestions on how we could reorganize the upset cells to reduce the downtime during die changeover and get to lower lot-size production."

After an active discussion and many trips to the flip chart to sketch alternative cell configurations, the group arrived at the initial system layout. The three manual upset cells, Figure 4–32, would be organized into a single cell, Figure 4–33, with three forges, two ovens, and two robots. One of the forges could be serviced by either robot, so that production would not stop when a die needed to be changed. Judging that most of the constructive ideas had been voiced, Bill took control of the meeting and said:

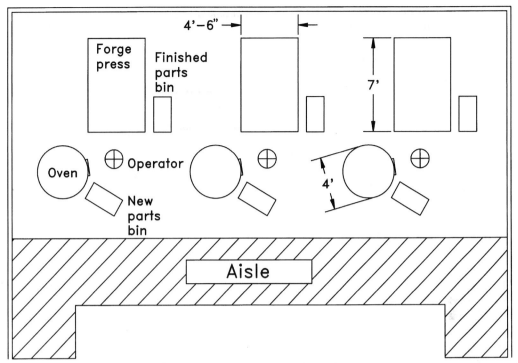

Figure 4–32 Manual Upset Forge Cells.

I like what I've heard today—we're on the right track. I have a list of to-do's for everyone that will carry us to the next step in the process. Before we start implementing the automation, however, we need to identify the performance measures that will be used to measure the project's success. Marci, could you do some research for the next meeting? In addition, I would like you to take the technical issues checklist in the folder and start identifying areas that need additional work. Mike, I would like you to work with me on the robot selection checklist to identify a robot for the cells.

Ted showed me the new Workspace 3 work-cell simulation software that we have in the design area. It should help us be more confident of our design and eliminate implementation surprises. Ted, could you give us an early look at equipment placement, cycle times, and robot work envelopes on the simulation package at next week's meeting? Jerry, please start looking into the modifications necessary for the ovens—especially, the automation of the internal parts table.

After some clarification, all members indicated that they would work on the areas listed. As the team prepared to leave, Bill said:

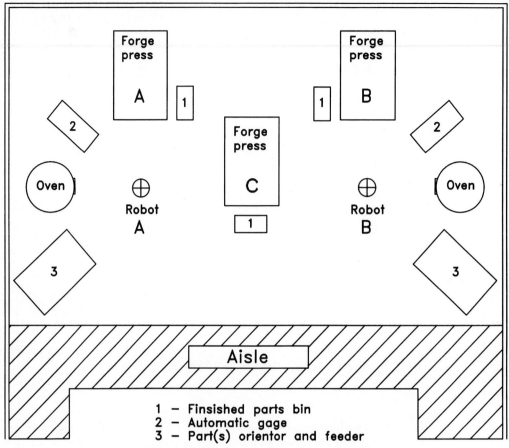

Figure 4–33 Proposed Automation Work Cell.

One last thing before we adjourn. Ted did some initial work on a gripper that could be used in either the extrusion or upset cells; the drawing (Figure 4–34) is in the folder. Ted, please give us a quick overview of the design.

Ted got out a large plot of the gripper drawing and said, "I chose stainless steel because it can withstand the high oven and part temperatures. The temperature rise in the gripper was another concern, and so I performed a stress and heat flow study on the design using finite-element analysis software. After some redesign, the heat rise in the robot tool plate was held to just 10 degrees over ambient temperature at current cycle times. I think the fingers attached to a standard parallel type pneumatic gripper will work fine on the cell layout we did today." ●

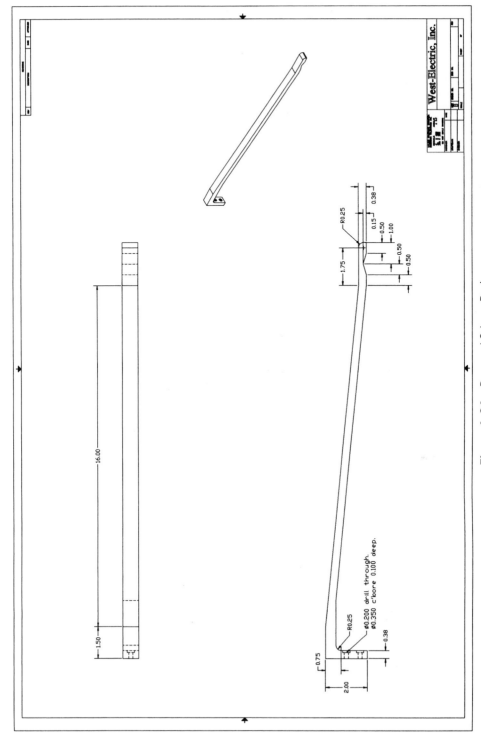

Figure 4–34 Proposed Gripper Design.

4-8 SUMMARY

The objective of this chapter was to introduce the different types of tooling that are currently available for use on robot arms. Grippers must be capable of holding the required part so that safety is assured under all operating conditions. In addition, grippers should conform to the lightest design using the simplest technique possible while holding and sensing the presence of the manufactured part.

In general, end-of-arm tooling groups include standard grippers, vacuum grippers, and special-purpose grippers and tools. Assembly fixtures require that special attention be paid to the problem of compliance. The alignment problem between mating parts can be solved either by active or by passive compliance. In active compliance, or force/torque sensing, the compensation for part alignment error occurs in the commands the controller sends to the arm. The adjusted coordinate values are generated as a result of signals received from the gripper indicating a force and torque resulted from an attempt to mate two parts. The second type of compliance is called passive and uses an RCC device. The required movement of the part to permit alignment and mating is produced in the RCC, which is mounted between the gripper and the robot wrist-mounting plate. The corrective movement results from forces acting on the part in the gripper as it is inserted into the assembly without perfect alignment.

Multiple end-effector systems have the advantage of permitting robot automation to be implemented in work cells involving complex operations. In addition, they permit increased use of expensive robots on a wider range of manufacturing problems. These advantages overcome the primary disadvantage, which is increased cost because of greater complexity.

The rate at which robots are introduced into new production situations is directly related to the rate at which the tooling develops. Many of the new applications require grippers that approximate the dexterity of the human hand. The development of a gripper with three or more fingers that can grasp a variety of irregularly shaped objects is critical for many future applications. Increased robot use is directly proportional to technical breakthroughs in end-of-arm tooling design.

QUESTIONS

1. What is end-of-arm tooling and what function does it serve?
2. What are the five characteristics that all end-of-arm tooling must satisfy?
3. What are the three categories for classifying end effectors, and what is included in each category?
4. Draw a classification tree that includes all the types of end-of-arm tooling described in Section 4–2.
5. What is the estimated cost for off-the-shelf and in-house designed tooling as a percentage of total system cost?

6. Write a procedure for calculating the lifting power of round vacuum cup grippers.
7. How does the complexity of the production job affect the intelligence level of the robot and tooling used for the application? How can that be changed?
8. What does the term *compliance* mean?
9. Describe the operation of an F/T sensor used for active compliance.
10. Describe the two methods used to achieve passive compliance.
11. Describe the three compliance conditions normally found in robot applications.
12. What is an RCC device?
13. What are the advantages and disadvantages of multiple end-effector systems?

PROBLEMS

1. Design a decision tree that will select the best end-of-arm tooling for a manufacturing problem. Use questions at each branch of the tree that require a *yes* or *no* response.
2. Write a computer program that will execute the design tree from project 1.
3. Write a computer program that will input the size and number of vacuum cups in a gripper along with the degree of vacuum achieved and output the lifting capability of the gripper.
4. The vacuum cups on the tooling in Figure 2–17 have a gripping area that is 1.5 inches in diameter. If the rear window glass weighs 17 pounds, what degree of vacuum would be required for a 50 percent safety factor? If the degree of vacuum is 40 percent and the safety factor is 60 percent, what is the maximum glass weight that can be lifted?
5. The gripper in Figure 4–11 is picking up 4-mil polyvinyl sheets that weigh 3 ounces per square foot. If the degree of vacuum behind the pickup surface is 10 percent, how many $\frac{1}{8}$ inch holes will be required per square foot for a 50 percent safety factor?

CASE PROJECTS

1. Is there a need to integrate upset forging and extrusion as Marci suggested? Does the proposed upset forging work cell automation eliminate the opportunity for integration with the extrusion process in the future? How could the integration be implemented?
2. Develop an initial layout for the automation of slug lubrication and extrusion using the existing extrusion machines.
3. What characteristics of good gripper design are satisfied by the gripper that Ted proposed for the work cell?

4. The mass properties analysis of Ted's gripper finger indicated that the volume of the material was 6.04 cubic inches. Calculate the weight of two fingers if they are made from stainless steel. What other materials are available that would provide the high-temperature operation characteristics but would contribute less weight?

5. Using the case study description of the W-E blade production line and any other resources available, complete the work-cell technical design checklist for the team's proposed upset forging automation systems.

6. Using the case study description of the W-E blade production line and any other resources available, complete the work-cell technical design checklist for the automation system developed in question 2.

7. Using case data, address and resolve as many of the robot checklist issues as possible for the upset forging automation system.

8. Using case data, answer as many of the robot checklist issues as possible for the work cell developed in problem 2.

Automation Sensors

5-1 INTRODUCTION

Even the most unskilled production worker can tell when a part has fallen to the floor or when a finished part is not ejected from a machine. The most sophisticated robot available today cannot perform these routine tasks without the help of sensors. Strategically mounted sensors provide the robot system with the same data that an operator gathers using the five human senses. For example, in Figure 5–1 a sensor checks for the presence of a part in the die of a forming press. A light source and light-sensitive receiver are positioned so that the beam of light from the source passes directly through the part in the die on its way to the receiver. Figure 5–1 shows this type of sensor detection system with the part in the die. A robot system would use a broken light beam or an *off* condition at the receiver to warn that the part had not been rejected and another part should not be loaded into the die.

Sensors are used in work cell for the following six basic reasons:

1. To detect a condition where an operator or some other human worker could be harmed by the robot or other manufacturing equipment
2. To detect a condition where the robot or other machines could be harmed by some other manufacturing equipment
3. To monitor the production system operation to ensure consistent product quality
4. To monitor the work-cell operation to detect and analyze system malfunctions

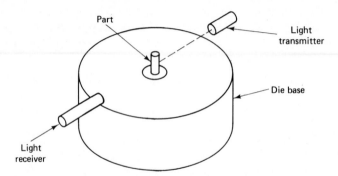

Figure 5–1 Part Ejector Failure Sensing.

5. To analyze production parts to determine the current level of product quality

6. To monitor production parts for identification, location, and orientation data that can be used by other production systems in the work cell

Sensors are grouped into two categories, called *contact* and *noncontact*. As the name implies, a contact sensor must physically touch an object before the sensor is activated. In contrast, a noncontact sensor measures its parameter of interest without touching the object of interest. In each of these two categories the sensors can have *discrete* or *analog* output signals.

Discrete Sensors

Discrete sensors have a single trigger point; for example, a temperature sensor changes output states when the temperature of the sensor crosses 68°F. A good example of this type of sensor is the thermostat that controls the heating systems in homes. A single temperature level is set, and the heating system cycles *on* and *off* at that temperature.

The two most distinctive characteristics of discrete sensors are that (1) a single-input condition triggers a change in sensor output state, and (2) the output of the sensor swings between two conditions, *on* and *off*. Many discrete sensors used in manufacturing signal the change in output states by the opening or closing of a set of electrical contacts. Other discrete sensors use voltage levels to indicate the *on* and *off* state of the output. Often the voltage levels are 0 and 24 volts dc or 0 and 115 volts ac for *off* and *on*, respectively. The trigger point for the sensor is usually adjustable so that the change in output states can be set to an input condition within a range dictated by the device.

Analog Sensors

In contrast to the discrete sensors, the analog type measures a range of input conditions and generates a range of output states. In analog temperature measurement, for example, the sensor system could produce an output that varies from 0 to 5 volts, or 4 to 20 milliamperes, when exposed to a temperature range of 0 to

100 degrees Centigrade (°C). A direct and linear relationship exists between the input condition and the output response. For example, 0°C produces 0 volts output, 100°C produces 5 volts output, and 50°C produces 2.5 volts output. Other sensors have inverse input and output relationships, and some types of sensors (eg, a thermocouple) produce a nonlinear output from a linear change in the input. Analog sensors include all sensors that measure process parameters, such as temperature, pressure, level, and flow. This type of sensor is found most frequently in process industries such as food, chemical, and petroleum.

The description of robot controllers in Chapter 2 indicated that robots use analog sensors to measure the position of axes and the rate of change of the joints as the robot moves. Although some analog sensors measure unique conditions in the work cell, more are found on the robot itself than in the automated cell where the robot works. The primary sensor used to automate manufacturing cells is the discrete type; therefore, the sensors discussed in this chapter will be primarily contact and noncontact discrete-type devices.

5-2 CONTACT SENSORS

Contact sensors include *limit switches* and all *tactile-sensing devices.* Limit switches, which have been used in flexible automation for many years, are reliable and easy devices to interface in an automated work-cell. Tactile sensing is in the development stage; a high level of research activity is occurring in both the industrial and university laboratory.

Discrete Devices

The most frequently used discrete contact sensor is the *limit switch.* The selection and application of limit switches requires an understanding of the *physical* and *electrical* properties of the devices along with their *operational characteristics.* As these properties and characteristics are described, refer to the device data sheet in Figure 5–2.

The *physical* properties of a limit switch include the fact that it is a mechanically actuated electrical switch consisting of a receptacle, switch body, operating head, and contacting device. On the Omron Model D4A data sheet (Figure 5–2b) two types of switch bodies, receptacles, and operating heads are shown. Note that the contacting device on one is a lever (D4A-☐☐01), and on the other it is a roller plunger (D4A-☐☐10). The receptacle provides the electrical connections for interfacing, whereas the switch body holds the switch and provides a mounting base for the various operating heads. The operating heads listed on the data sheet will mount to a standard switch body. The rotary type requires the contacting devices to rotate to activate the switch, whereas the two types of plunger heads require a linear movement of the plunger to cause switching of the contacts. The wobble-lever type (D4A-☐☐12) has a long trip rod attached to the head that will trip the switch if the rod is moved away from the center position in any direction. As the

OMRON
GENERAL-PURPOSE LIMIT SWITCH

Cat. No. C03-E3-2

Model **D4A**

Heavy Duty Plug-in Limit Switch— Watertight and Oiltight

■ FEATURES

- Convenient front-mount plug-in construction.
- High-reliability design.
- Wide variety of operating heads.
- Meets NEMA types 1, 2, 3, 3R, 4, 4X, 5, 6, 12 and 13.

- Certainty of dual Viton® sealing.
- Easy installation and maintenance.
- Reduces inventory.
- Low temperature type (—40 to +158° F) is also available.

■ AVAILABLE TYPES

- SWITCH UNITS (D4A-☐☐☐: Add the appropriate code when placing your order, e.g., D4A-1101.)

Operating head	Circuit	SPDT DOUBLE BREAK				DPDT DOUBLE BREAK	
	Indicator lamp	Without lamp		With lamp		Without lamp	
	Contact rating	Pilot duty	Electronic duty	Pilot duty	Electronic duty	Pilot duty	Electronic duty
Side* rotary type	Standard	-1101	-1201	-1301	-1401	-2501	-2601
	High precision	-1102	-1202	-1302	-1402	-2502	-2602
	Low torque	-1103	-1203	-1303	-1403	-2503	-2603
	High precision/Low torque	-1104	-1204	-1304	-1404	-2504	-2604
	Maintained	-1105	-1205	-1305	-1405	-2505	-2605
	Sequential operating					-2717	-2817
	Center neutral operating					-2918	-2018
Side plunger type	Plain	-1106	-1206	-1306	-1406	-2506	-2606
	Vertical roller	-1107-V	-1207-V	-1307-V	-1407-V	-2507-V	-2607-V
	Horizontal roller	-1107-H	-1207-H	-1307-H	-1407-H	-2507-H	-2607-H
	Adjustable	-1108	-1208	-1308	-1408	-2508	-2608
Top plunger type	Plain	-1109	-1209	-1309	-1409	-2509	-2609
	Roller	-1110	-1210	-1310	-1410	-2510	-2610
	Adjustable	-1111	-1211	-1311	-1411	-2511	-2611
Wobble lever type	Spring wire	-1112	-1212	-1312	-1412	-2512	-2612
	Plastic rod	-1114	-1214	-1314	-1414	-2514	-2614
	Cat whisker	-1115	-1215	-1315	-1415	-2515	-2615
	Coil spring	-1116	-1216	-1316	-1416	-2516	-2616

- **LEVERS FOR SIDE ROTARY HEAD**

Standard-roller front mounted	.75 dia./.31 width	D4A-A00
	.68 dia./.59 width	D4A-B06
Standard-roller back mounted		D4A-A10
Offset-roller front mounted		D4A-A20
Offset-roller back mounted		D4A-A30
Adjustable-roller		D4A-C00
Rod-adjustable		D4A-D00
Fork-rollers on opposite side		D4A-E00
Fork-rollers on opposite side		D4A-E10
Fork-rollers on same side		D4A-E20

NOTES: 1. * When placing your order for a side rotary type switch, order the lever with the switch.
2. With the standard types, only the switch body and shaft parts are Viton® seated and all other parts are sealed with nitrile-butadiene rubber (NBR). If the Viton® sealed type (with all parts sealed with Viton® rubber) is required, add suffix code "F" to the desired part number shown on the left (e.g., D4A-1101-F).
3. If the low temperature type is required, add suffix code "T" to the part number shown on the left (e.g., D4A-2501-T).
4. To order switch bodies, receptacles and operating heads separately, refer to "REPLACEMENT PARTS."

— OMRON —

■ SPECIFICATIONS

- **RATINGS**
- Pilot duty type (NEMA A600/B600)

Circuit	Rated voltage	Current (A)		Voltamperes (VA)	
		Make	Break	Make	Break
SPDT DOUBLE BREAK	120 VAC	60	6	7,200	720
	240 VAC	30	3		
	480 VAC	15	1.5		
	600 VAC	12	1.2		
DPDT DOUBLE BREAK	120 VAC	30	3	3,600	360
	240 VAC	15	1.5		
	480 VAC	7.5	0.75		
	600 VAC	6	0.6		

- **Electronic duty type**
This type is ideal for use in the load range (Zones 1 through 3) shown below.

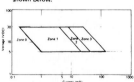

Approved by Standards
UL (File No. E68575)
(Control No. Ind. Cont. Eq. 298Z)
Rating NEMA A600
600V AC MAX PILOT DUTY*
CSA (File No. LR45746)
Rating NEMA A600
600V AC MAX PILOT DUTY*

* PILOT DUTY is one of the rating classifications specified for "UL917 clock-operated switches" and refers to large inductive loads with a power factor of 0.35 or less.

- **CHARACTERISTICS**

Operating speed	.04'' to 6.56''/sec (at side rotary type)
Operating frequency	Mechanically: 300 operations/min. Electrically: 30 operations/min.
Contact resistance	25mΩ max. (initial)
Insulation resistance	100MΩ min. (at 500 VDC) between each terminal and non-current-carrying part and ground
Temperature rising	30 degree max.
Dielectric strength	1,000 VAC, 50/60Hz for 1 minute between non-continuous terminals; 2,200 VAC, 50/60Hz for 1 minute between each terminal and non-current-carrying metal part and between each terminal and ground
Vibration	Malfunction durability: 10 to 55Hz, .06'' double amplitude
Shock	Mechanical durability: Approx. 100G's Malfunction durability: Side rotary type: Approx. 60G's Non side rotary type: Approx. 30G's
Ambient temperature	Operating: D4A-11☐☐/-12☐☐ series: +14 to +248° F (D4A-1103/-1203: +23 to +248° F) D4A-13☐☐/-14☐☐ series: +14 to +176° F D4A-25☐☐/-26☐☐ series: +14 to +194° F D4A-☐☐☐☐-T series: —40 to 158° F
Humidity	95% RH max.
Degree of protection	NEMA	Types 1, 2, 3, 3R, 4, 4X, 5, 6, 12, 13
	IEC	IP-67
	JIS	Immersion-proof type
Service life	Mechanically: 50,000,000 operations min.
Weight	Approx. 20.6 oz. (D4A-1101)

Figure 5–2 Limit Switches: (a) General Specifications; (b) Electrical Contacts and Operating Specifications; (c) Trip Mechanisms. (Courtesy of Omron Electronics, Inc.)

● CONTACT CONFIGURATION

SPDT DOUBLE BREAK
(Must be same polarity)

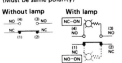

Without lamp　With lamp

DPDT DOUBLE BREAK
(Same polarity each pole)

Sequential operating type:
Pole 1 operates first
Pole 2 operates second
Either CW or CCW or Both

Center neutral operating type:
Pole 1 operates CW
Pole 2 operates CCW

● OPERATING CHARACTERISTICS

Type	D4A-□□01	D4A-□□02	D4A-□□03	D4A-□□04	D4A-□□05	D4A-□□06	D4A-□□07	D4A-□□08	D4A-□□09
OF oz. max.	3.5 lb-in	3.5 lb-in	1.7 lb-in	1.7 lb-in	3.5 lb-in	95.2	95.2	95.2	63.5
RF oz. min.	.4 lb-in	.4 lb-in	.3 lb-in	.3 lb-in	—	31.7	31.7	31.7	17.6
PT in. max.	15°	7°	15°	7°	65°	.094	.094	.094	.063
OT in. min.	60°	68°	60°	68°	20°	.201	.201	.201	.201
MD in. max.	5°	4°	5°	4°	35°	.023	.023	.023	.016
	(7°)	(6°)	(7°)	(6°)	(35°)	(.039)	(.039)	(.039)	(.039)
OP in.	—	—	—	—	—	1.339±.031	1.732±.031	1.614 to 1.870	1.811±.031

Type	D4A-□□10	D4A-□□11	D4A-□□12	D4A-□□14	D4A-□□15	D4A-□□16	D4A-□□17	D4A-□□18
OF oz. max.	63.5	63.5	3.5	5.3	5.3	5.3	3.5 lb-in	3.5 lb-in
RF oz. min.	17.6	17.6	—	—	—	—	.2 lb-in	.2 lb-in
PT in. max.	.063	.063	3.937 Radius	1.378 Radius	1.969 Radius	1.969 Radius	1st Pole 10° 2nd Pole 18°	15°
OT in. min.	.201	.201	—	—	—	—	57°	60°
MD in. max.	.016	.016	—	—	—	—	Each Pole 5°	5°
	(.039)	(.039)						
OP in.	2.205±.031	2.185 to 2.441	—	—	—	—	—	—

NOTE: Figure in parenthesis applies to the DPDT double break circuit type (D4A-2□□□).

■ DIMENSIONS (Unit: inch)

NOTES: 1. Unless otherwise specified, a tolerance of ±.016″ applies to all dimensions.
2. All dimensions shown here are for reference only.

SWITCH UNITS

● D4A-□□01 ~ D4A-□□05
　D4A-□□17 ~ D4A-□□18

● D4A-□□06

● D4A-□□07-V

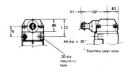

● D4A-□□07-H

● D4A-□□08

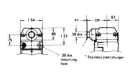

● D4A-□□09

● D4A-□□10

● D4A-□□11

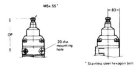

● D4A-□□12

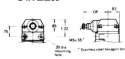

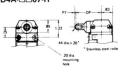

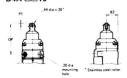

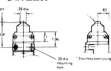

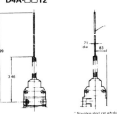

Figure 5–2 *(continued)*

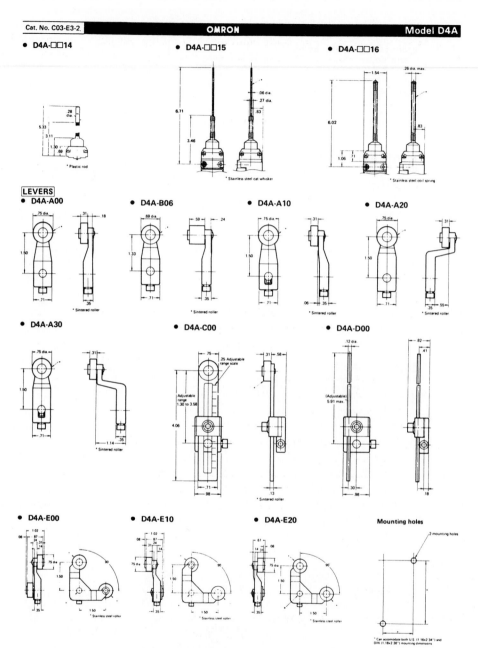

Figure 5–2   *(continued)*

data indicate, the different head types are available with various contacting devices. For example, both plain and adjustable plungers are provided, and the wobble-lever type has four different contacting rods. Some of the many levers available for rotary-type heads include standard roller, offset roller, adjustable roller, rod adjustable, and fork roller. Illustrations of the different types of contactors are provided in Figure 5–2c.

The *electrical* properties include the contact ratings for current and voltage and the contact configuration. The contact ratings are broken into two categories called *pilot duty* and *electronic duty* (Figure 5–2a). Pilot duty assumes switching voltages of 120 to 600 volts ac with currents from 60 to 0.6 amperes, respectively. These would be normal service conditions for controlling motors, motor starters, lamps, and other high-power devices. The ratings table under "Specifications" on the D4A data sheet classifies these currents depending on operating voltage and contact closure (make) or contact opening (break) conditions.

Electronic duty covers the operation at low values of voltage, 5 to 30 volts, and low values of current, 0.2 to 100 milliamperes. The measure of a switch's ability to handle low currents is called the *dry circuit rating*. An input module of a programmable logic controller requires less than 150 milliamperes when 28 volts are applied. Therefore, limit switches interfaced directly to programmable logic controllers should have contacts with good low-current switching characteristics. The contact configurations available in the D4A model are illustrated at the top of Figure 5–2b. Both are double-break–type switches, which means the contactor leaves both of the normally closed (NC in Figure 5–2b) contacts when the switch is actuated. In the two-pole model the operation is either sequential or center-neutral type. The sequential type has pole 1 closing first, followed by pole 2 as the switch is actuated. In the center-neutral type, pole 1 operates during clockwise rotation of the lever and pole 2 during counterclockwise rotation.

The *operational characteristics* of the model D4A switch include everything from the operating speed to the device weight. Review the characteristics listed on the data sheet and note the excellent service life of these devices. The definition of various switch movements and the forces required to produce them are also important operational characteristics common to all switch models. Figure 5–3 shows a listing and illustration of the movements and forces for the plunger and lever-type switch. A lamp to indicate if the switch has power applied is available as an option.

Artificial Skin

Effective tactile sensors and systems are needed for robot grippers used in assembly applications. Research on small and medium-sized parts-assembly tasks indicates that the development of a basic form of artificial skin would permit robots to perform about 50 percent of the generic tasks found in industrial assembly. Currently, most robot interaction with sensors is of the open-loop type. For example, gripping devices close on all parts with the same force. A signal indicating that the gripper is closed is generated by the sensor on the robot end-of-arm tooling, but feedback to the robot controller of gripping characteristics, such as closing force, is not provided. In grippers with artificial skin, the fingers are closed

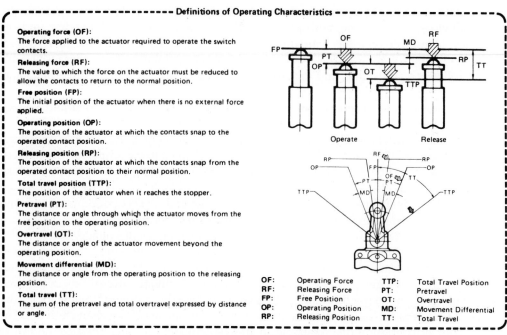

Figure 5–3 Definition of Switch Movement. (Courtesy of Omron Electronics Inc.)

under servo control with finger pressure and other tactile parameters measured. There is an important distinction between *tactile sensing* and *simple touch*. Simple touch includes simple contact or force sensing at one or just a few points on the gripper surface. For example, the gripper in Figure 5–4 uses two switches to determine that the part is centered in the jaws. The switches provide binary data in the form of either an *on* or *off* signal depending on the location of the part. If the switches were replaced with analog pressure pads, the data would be continuously variable, and the signal value would become a function of the pressure applied by the fingers. The sensing would still be classified as simple touch, however, because only two contact points were monitored.

Tactile sensing requires a group of sensors arranged in a rectangular or square pattern called an array. Figure 5–5 shows gripper fingers with an 8 × 8 tactile-sensing array. The array has sixty-four sensing elements, each capable of measuring the continuously variable force applied to the element. A tactile sensor interfaced to an intelligent controller can determine the shape, texture, position, orientation, deformation, center of mass, and presence of torque and slippage of any object held. The array output pictured in Figure 5–5 shows the results of holding a triangular-shaped object in the gripper.

The three steps required for intelligent acquisition of parts by either a robot or a human operator are first *vision*, followed by *proximity sensing*, and finally *tactile sensing*. For a robot system to retrieve randomly placed parts or to assemble

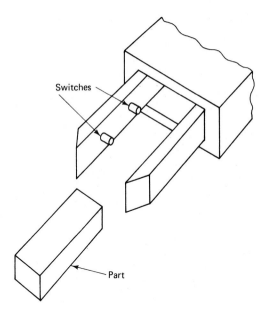

Switches

Part

Figure 5–4   Simple Touch-Sensing Gripper.

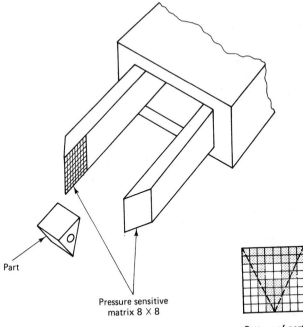

Part

Pressure sensitive
matrix 8 × 8

Pattern of part

▨ = Pressure sensed

☐ = No pressure

Figure 5–5   Tactile-Sensing Array.

Classification		HIGH-FREQUENCY OSCILLATION TYPE			
Model		TL-X (Compact type)	TL-X (Plug-in type)	TL-XD	E2P
Features		Cylindrical type proximity sensor boasting ultra small size and high performance	Heavy duty proximity sensor with low operating current and integral receptacle	Cylindrical type proximity sensor with DC-2 wire system	Heavy duty proximity sensor with low operating current and integral receptacle
Appearance & dimensions		TL-X1□□ TL-X18MY□	TL-X5Y	TL-XD(B)5 TL-XD(B)15M	
Sensing distance		1, 2, 5, 10, 18mm	5, 10mm	2, 4, 5, 8, 10, 15mm	15mm
Rated voltage	DC switching type	10 to 40 VDC	—	8 to 40 VDC	—
	AC switching type	45 to 260 VAC	90 to 140 VAC	—	90 to 140 VAC
Operating current	DC switching type	10mA max.	—	1.5mA max.	—
	AC switching type	2mA max.	1.7mA max.	—	1.7mA max.
Materials sensed		All metals	All metals	All metals	All metals
Response frequency or time	DC switching type	1kHz min.	—	800Hz min.	—
	AC switching type	20Hz min.	12.5Hz min.	—	12Hz min.
Control output (switching capacity)	DC switching type	200mA max.	—	3 to 200mA	—
	AC switching type	5 to 200mA	.5 to 700mA	—	5 to 700mA
Degree of protection	JIS C0920	—	Water-resistant type	—	Water-resistant type
	IEC 144	IP67	IP67	IP67	IP67
	NEMA	Types 1, 4, 6, 12, 13	Types 1, 2, 4, 6, 12, 13	Types 1, 4, 6, 12, 13	Types 1, 2, 4, 6, 12, 13
Ambient temperature		Operating: −40 to +85°C	Operating: −25 to +70°C	Operating: −25 to +70°C	Operating: −25 to +70°C
Approved standards		(UL)	(UL)	—	(UL)
Page		287	293	297	301

Figure 5–6 Proximity Sensors. (Courtesy of Omron Electronics Inc.)

products at any scale automatically, further development of all three of these sensing areas is required. In addition, if machines are to develop tactile sensing such as that found in humans, then a second element must be incorporated. To the human faculty of skin sensing is added *haptic perception,* which is sensing information that comes from the joints and muscles. Thus robot software must combine the data from the artificial skin with the joint torque information if machines are to achieve what humans call "touch."

5-3 NONCONTACT SENSORS

As the name implies, noncontact sensors measure work-cell conditions without physically touching the part. In robot work cells the most frequently used noncon-

HIGH-FREQUENCY OSCILLATION TYPE	ELECTROSTATIC CAPACITANCE TYPE	HIGH-FREQUENCY OSCILLATION TYPE		
TL-L	E2K	TL-M	TL-N/H/F	TL-YS10
High-performance proximity sensor with 50mm sensing distance & LED for short-circuit indication	Capacitive type senses nonmetalic objects.	Basic switch housing model	High-frequency oscillation type with easy operation monitoring	Direct AC load switching type with easy-to-see operation indicator lamp
TL-LP50			TL-N5ME□ TL-H10ME□ TL-F20ME□	
50mm	3 to 25mm	2, 5mm	5, 6, 8, 10, 12, 20mm	10mm
10 to 30 VDC	10 to 40 VDC	10 to 30 VDC	10 to 30 VDC	—
90 to 250 VAC	90 to 250 VAC	90 to 250 VAC	90 to 250 VAC	90 to 130, 180 to 260 VAC
10mA max.	10mA max. (at 12 VDC) 15mA max. (at 24 VDC)	15mA max. (at 24 VDC)	8 mA max. (at 12 VDC) 15mA max. (at 24 VDC)	—
—	1mA max. (at 100 VAC 50/60Hz) 2mA max. (at 200 VAC 50/60Hz)	1.5mA max. (at 100 VAC 50/60Hz) 3.0mA max. (at 200 VAC 50/60Hz)	2mA max.	5mA max.
All metals	Non-metals, Metals	All metals	All metals	All metals
15msec max.	70Hz min.	500Hz min.	500Hz min.	—
25msec max.	10Hz min.	20Hz min.	10Hz min.	10msec max.
200mA max.	200mA max.	100mA max. (at 12 VDC) 200mA max. (at 24 VDC)	200mA max.	—
10 to 200mA	5 to 200mA	10 to 200mA	5 to 200mA	10 to 700mA
	Water-resistant type		—	Water-resistant type
IP67	IP66	IP67	IP67	IP66
Types 1, 4, 6, 12, 13	Types 1, 4, 12, 13	Types 1, 4, 12, 13	Types 1, 4, 6, 12, 13	Types 1, 4, 12, 13
Operating: −25 to +70°C	Operating: −25 to +70°C	Operating: −25 to +70°C	Operating: −25 to +70°C	Operating: −25 to +70°C
५	५	—	—	—
305	309	313	317	321

Figure 5–6 *(continued)*

tact sensors are *proximity* and *photoelectric* devices and *vision* systems. All three types of sensors are available from many commercial vendors. Proximity and photoelectric sensors are discussed in the following sections, and vision systems are described in Chapter 6.

Proximity Sensors

Proximity sensors detect the presence of a part when the part comes within a specified range of the sensor. Proximity sensors are available in four package shapes and several different sizes. The package shapes are *cylindrical, rectangular, through-head type,* and *grooved-head type.* The shape of the part to be sensed and sensing application dictate the type of package shape for best operation. Figure 5–6 shows the basic sensors available from Omron. The sensors are mounted to supporting brackets using either a threaded part of the sensor body or mounting holes. The operating parameters for proximity sensors are defined as follows:

- *Reference plane:* The plane of reference on the proximity sensor from which all measurements are made
- *Reference axis:* An axis through the sensor from which measurements are made
- *Standard object:* A definition of the object to be sensed in terms of a specified shape, size, and material composition
- *Sensing distance:* The distance from the reference plane to the standard object that causes the output of the sensor to change to the *on* state
- *Vertical sensing distance:* The sensing distance measured by bringing the standard object toward the reference plane with the standard object normal to and centered about the reference axis. (See Figure 5–7.)
- *Horizontal sensing distance:* The sensing distance measured by bringing the standard object along a plane parallel with and at a fixed distance from the reference plane. (See Figure 5–8.)
- *Resetting distance:* The distance from the reference plane to the standard object that causes the output of the sensor to change from *on* to *off* as the standard object is withdrawn from the sensor. (See Figures 5–7 and 5–8.)
- *Differential distance:* A measure of the hysteresis present in the system. The difference between the resetting distance and the sensing distance for the type of sensor used. (See Figures 5–7 and 5–8.)

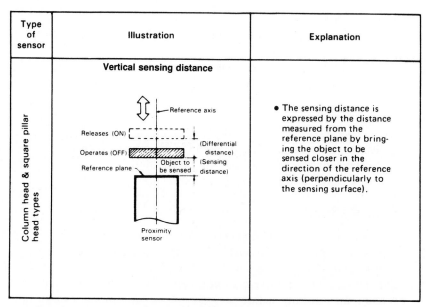

Type of sensor	Illustration	Explanation
Column head & square pillar head types	**Vertical sensing distance**	• The sensing distance is expressed by the distance measured from the reference plane by bringing the object to be sensed closer in the direction of the reference axis (perpendicularly to the sensing surface).

Figure 5–7   Vertical Sensing Distance. (Courtesy of Omron Electronics Inc.)

- *Setting distance:* The maximum sensing distance when worst-case ambient temperature and supply voltage variations are assumed. (See Figures 5–7 and 5–8.)
- *Response time:* The time required for the output to change states after the standard object's position triggered a change of state.
- *Frequency response:* The maximum rate at which standard objects can cause the output to change states.
- *Leakage current:* The maximum *off* current that will flow from the output terminals with the output stage in the *off* state.
- *Temperature variation:* The variation in sensing distance as a result of variations in ambient temperature.
- *Voltage variations:* The variation in sensing distance as a result of variations in supply voltage.

The selection and the application of proximity devices requires an understanding of the *physical* and *electrical* properties of the sensors along with their *operational characteristics*.

All models have solid-state circuits enclosed in the sensing head to generate the fields necessary for remote sensing of objects and production of output signals. Most sensors are designed to detect metal objects, whereas a few models are available that can detect both metallic and nonmetallic parts. The sensor that detects metal only uses a high-frequency electromagnetic field for detecting part presence, whereas the nonmetallic type uses electrostatic capacitance for the detecting mechanism.

Type of sensor	Illustration	Explanation
Column head & square pillar head types	**Horizontal sensing distance** (Sensing distance) (Differential distance) Object to be sensed — Releases (OFF) — Operates (ON) — Reference axis — Reference plane Proximity sensor	• The sensing distance is expressed by the distance measured from the reference axis by moving the object to be sensed parallel to the reference plane (i.e., sensing surface). This distance can be expressed as the locus of operating point, since it varies with the passing position of the object (distance from the reference plane).

Figure 5–8   Horizontal Sensing Distance. (Courtesy of Omron Electronics Inc.)

Some manufacturers provide three different output circuits for interfacing to other work-cell hardware. The output circuits in Figure 5–9 include NPN and PNP transistor outputs in the dc switching type and an ac switching-type output. (See Figure 5–9.) Note that the sensor shares the power source with the external interface, and the sensor output connects either to a load or to an external transistor circuit. The NPN type of sensor output has a current-sinking characteristic. In current sinking, the output transistor of the activated sensor provides a low resistance path to the *black* lead, and so the sensor output is approximately zero volts. Under these conditions current would flow from the external interface device into the sensor output. In contrast, the PNP output circuit has a current-sourcing characteristic. When active, the sensor output transistor provides a low resistance path to the *red* lead so that the sensor output is approximately equal to the supply voltage. Under these conditions the sensor is supplying current to the external interface device. The last type of output in Figure 5–9 is an ac switching type. In this type of output, the sensor has two output leads (*white and black*) that act as a switch for a source and load in series. The interface between sensors and work-cell control devices is described in greater detail in the next chapter.

The Omron models also offer the option of normally closed (NC) operation or normally open (NO) action at the output. The wave forms in Figure 5–9 illustrate the NC and NO operation of sensors when they are close to an object (*present*) and when they are away (*absent*). When the wave form is at the *operates* level, the sensor has current flow in the output lead; however, when the wave form is at the *release* level, no current is flowing. Study the charts until the concept of NC and NO operation is clear.

Most models permit a wide range of voltages for operating power. The model TL-X by Omron (Figure 5–6), for example, allows the supply power on dc output models to range between 10 and 40 volts dc unregulated, and the ac output devices can use 45 to 260 volts ac. The output can be interfaced directly to a robot controller, programmable logic controller, or be connected to a sensor controller provided by the sensor manufacturer. For example, the sensor controller available from Omron is powered from 110 volts ac and provides 12 or 24 volts dc (depending on the model) to power the sensor. The sensor controller also provides relay contact outputs for control of other work-cell machines, plus an inverting switch to change between NC or NO operation. The sensor controller, pictured in Figure 5–10, also has two sensor input connections so that sensor outputs can be logically combined using either the AND or OR operators.

The application of proximity sensors is straightforward. First, select the sensor model that corresponds to the characteristics of the parts to be sensed. Second, mount the sensor so that the object to be detected passes within the sensing distance of the device. Third, apply power either through a sensor controller or from an external source. Fourth, connect the relay contacts in the sensor controller to a robot or programmable logic controller that will use the sensor signal for machine control. Although this operation is simple, numerous design decisions must be made before each of the four steps is taken. In addition, the actual wiring diagram will vary depending on the sensor and controller selected.

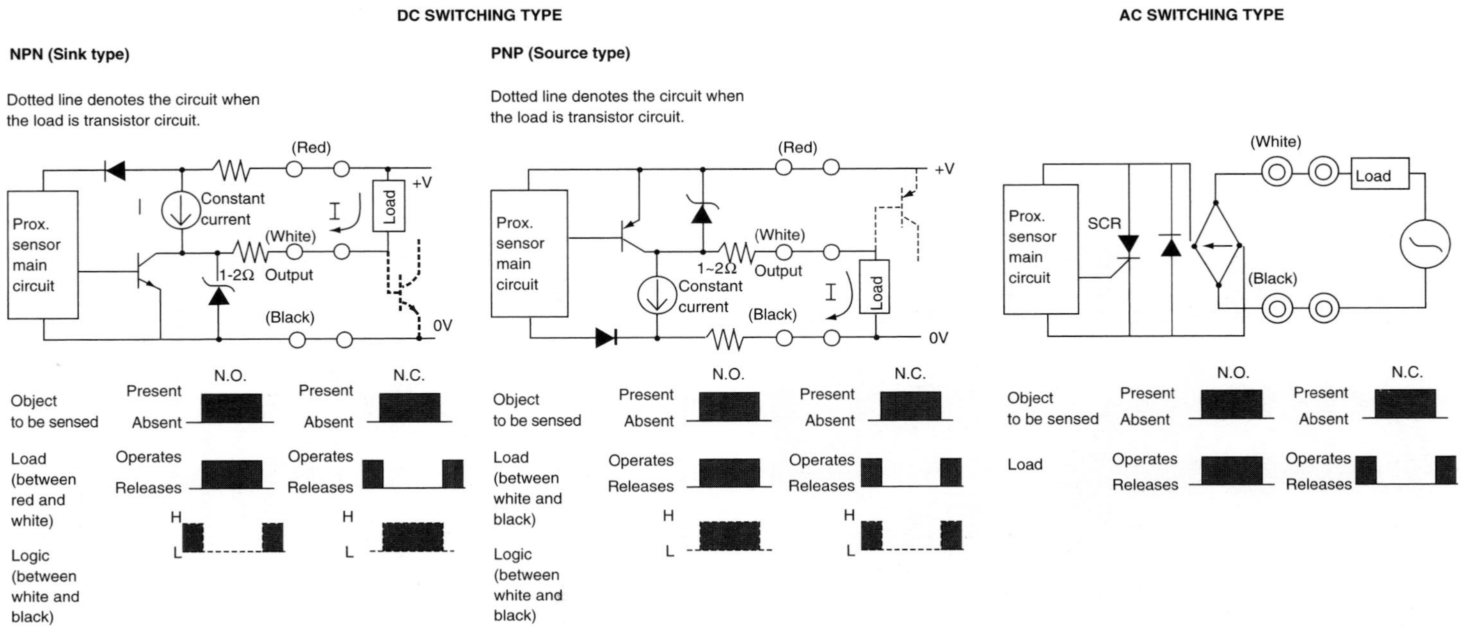

AC SWITCHING TYPE

DC SWITCHING TYPE

NPN (Sink type)

Dotted line denotes the circuit when the load is transistor circuit.

PNP (Source type)

Dotted line denotes the circuit when the load is transistor circuit.

Prox. sensor main circuit

Load

Constant current

Output

Constant current

(Red) (White) (Black)

+V

0V

1–2Ω

	N.O.		N.C.	
Object to be sensed	Present	Absent	Present	Absent	
Load (between red and white)	Operates	Releases	Operates	Releases	
Logic (between white and black)	H L		H L		

Object to be sensed: Present / Absent

Load (between white and black): Operates / Releases

Logic (between white and black): H L

Prox. sensor main circuit

SCR

Load

(White)

(Black)

Object to be sensed: Present / Absent

Load: Operates / Releases

Figure 5–9 Sensor Output Circuits.

165

Figure 5–10 Sensor Controller.
(Courtesy of Omron Electronics Inc.)

Sensing distance is affected by many variables including the following six:

1. Type and model of sensor
2. Material being sensed
3. Path the object uses to trigger the sensor
4. Ambient temperature variation
5. Supply voltage variation
6. Proximity of other objects and other sensors

For example, an Omron model TL-X10 used to detect a brass plate 50 mm along one side, using a vertical approach, will have a sensing distance of 4 mm. That distance could vary by ± 17.5 percent depending on variations in the temperature and supply voltage. In addition, if another proximity sensor must be mounted closely to this unit, then the second sensor must use a different operating frequency to avoid interference.

Interfacing the power and output signals requires an equally large number of design considerations. One factor affecting the sensor selection process is the input characteristics of the machine to which the sensor is connected. A programmable logic controller or robot controller, for example, will require a low-voltage type of sensor output circuit (usually 0 to 5 or 0 to 28 volts dc), and a motor starter control relay or contactor would require a high-power–type sensor output (usually 0 to 110

Term	Definition
Sensing distance	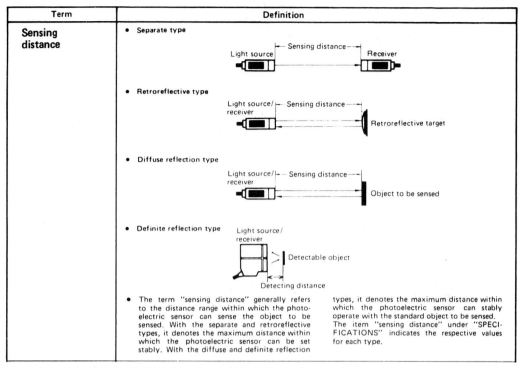 • Separate type • Retroreflective type • Diffuse reflection type • Definite reflection type • The term "sensing distance" generally refers to the distance range within which the photoelectric sensor can sense the object to be sensed. With the separate and retroreflective types, it denotes the maximum distance within which the photoelectric sensor can be set stably. With the diffuse and definite reflection types, it denotes the maximum distance within which the photoelectric sensor can stably operate with the standard object to be sensed. The item "sensing distance" under "SPECIFICATIONS" indicates the respective values for each type.

Figure 5–11 Four Types of Photoelectric Sensors. (Courtesy of Omron Electronics Inc.)

volts ac). The decision to connect the sensor directly to the machine control versus using a sensor controller will be affected by the availability of power for the sensor and the need for logical operations with two or more sensors.

Photoelectric Sensors

Photoelectric sensors detect the presence of an object or part when the part either breaks a light beam or reflects a beam of light to a receiver. The different types, pictured in Figure 5–11, are defined as follows:

- *Separate type:* The sensor system includes two separate devices: a light source to produce a beam of light and a receiving device to sense the presence of the light beam.
- *Retroreflective type:* The sensor system includes two separate devices: a sensor with both a light source and receiver in the same case, and a retroreflective target, which is a highly reflective surface. The light beam leaves the sensor, bounces off the target, then returns to the receiver. The sensor is triggered when a part breaks either the outbound or returning beam.

- *Diffuse reflective type:* The sensor has a light source and receiver built into the same case. Stable operation assumes that the part to be detected will return sufficient diffused light to trigger the receiver when the part is in range, and that the background equipment will not return enough light to trigger the receiver when the part is out of range. The light beam is diffused back to the receiver by using the natural reflective characteristics of the part's surface.

- *Definitive reflective type:* This sensor works like a combination of the retrore-flective and diffuse reflective types. The light source and receiver are both located in the same enclosure, and the beam uses the part's surface to reflect light back to the receiver. The light reflected must be a definitive beam because diffused light alone will not be sufficient to activate the receiver. The angle of the light leaving the sensor is adjusted to improve sensitivity and establish a specific distance at detection.

Several operating parameters describe the operation of the sensors and they are defined as follows:

- *Sensing distance:* For separate and retroreflective types it is the maximum distance between the light source and the receiver, or light source and target, that produces stable operation.

- *Operating distance:* For the reflective type it is the distance from the sensor to the part that causes the output of the sensor to change to the *on* state.

- *Resetting distance:* For the reflective type it is the distance from the sensor to the part that causes the output to change from *on* to *off* as the part is withdrawn from the sensor.

- *Differential distance:* A measure of the hysteresis present in the system. The hysteresis is the difference between the resetting distance and the operating distance for the type of reflective sensor used.

- *Optical axis:* The axis passing through the sensor along which the light beam is generated or received.

- *Direction angle:* For separate and retroreflective-type sensors it is the maximum angle through which the optical axis can move and still provide stable operation.

- *Dark on operation:* For all types of sensors it means that the output will be active, or *on,* when no light is received by the receiver.

- *Light on operation:* For all types of sensors it means that the output will be active, or *on,* when light is received by the receiver.

- *Response time:* The time required for the output to change states after the part breaks the beam or reflects the light to the receiver. Response time can be either operate time (time to turn the output *on*) or reset time (time to turn the output *off*).

The selection and application of photoelectric sensors requires an understanding of the *physical, electrical,* and *optical properties* along with the *operational characteristics* of the different types of sensors. The various types of sensors available from Omron are illustrated in Figure 5–12 on pp. 169–170.

Classification	AMPLIFIER SELF-CONTAINED TYPE			
Model	E3A	E3B	E3N	E3S
Features	Slim styled photoelectric sensor with built-in amplifier	Slim styled photoelectric sensor with built-in relay output in rigid die-cast case	Built-in amplifer type with convenient bicolor indicator	Miniature water-resistant photoelectric sensor with built-in amplifier
Appearance & dimensions	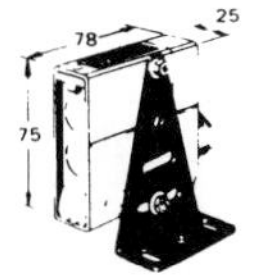		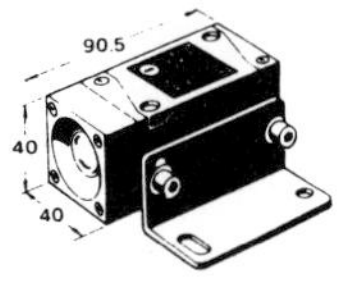	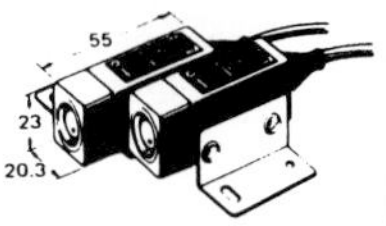
Method of sensing	Separate type Retroreflective type Diffuse reflection type	Separate type Retroreflective type Diffuse reflection type	Separate type Retroreflective type Diffuse reflection type	Separate type Retroreflective type Diffuse reflection type
Rated voltage	90 to 250 VAC, 50/60Hz	120 VAC, 50/60Hz	12 VDC −10% to 24 VDC +10%	12 VDC −10% to 24 VDC +10%
Operating current	2, 3, 4.5V A max.	4V A max.	60 or 80mA max.	40mA or 50mA max.
Sensing distance	0.1 to 5m	1, 5 or 10m	50cm, 2, 5, 10 or 30m	1.2 ±0.2, 5 or 20cm 0.1 to 1, 1 or 3m
Object to be sensed	Transparent and opaque materials	Transparent and opaque materials	Transparent and opaque materials	Transparent and opaque materials
Control output	Contact output (SPDT) 250 VAC 1A Solid-state output (SCR) Switching capacity: 250 VAC, 200mA max.	Pilot duty 120/240 VAC 180VA, 5A 120/240 VAC resistive	Solid-state output: Output (source) current: 1.5 to 3mA Load (sink) current: 200mA max.	Solid-state output: Output (source) current: 1.5 to 3mA Load (sink) current: 80mA max.
Response time	Comtact output: 60msec max. SCR output: 30msec max.	30msec max.	5msec max.	5msec max.
Ambient temperature	Operating: −10 to +55°C	Operating: −10 to +55°C	Operating: −25 to +55°C	Operating: −25 to +55°C
Degree of protection — JIS (C 0920)	Water-resistant type	Water-resistant type	Drip-proof type	Water-resistant type
Degree of protection — IEC144	IP66	IP66	IP65	IP62 or IP66
Degree of protection — NEMA	Types 1, 4, 4X, 12	Types 1, 4, 4X, 12, 13	Types 1, 2, 12	Types 1, 2 or 1, 4, 4X, 12
Approved standards	—	(UL) (SP)	(UL) •	(UL) •
Page	331	335	339	345

NOTE: * DC switching type only.

Figure 5–12 Photoelectric Sensors (continues on p. 170). (Courtesy of Omron Electronics Inc.)

The primary difference among the four types of photoelectric sensors is in their range, that is, distance to the detected part. The separate-type sensor system permits the light source and receiver to be separated by up to 30 meters for some models, so that detection can be anywhere along the 30-meter length. The retroreflective type has separation distances between the sensor and target from 1 to 5 meters. The diffused-light type has a detection distance that is less than 1 meter and typically in the 5- to 50-centimeter range. Last, the definite reflective type has the shortest detection distance—typically from 5 to 25 centimeters. Although the range is small for the definite reflective type, the ability to operate with high ambient light conditions makes this type of sensor superior to the diffuse reflective type under those operating conditions.

The sensor unit consists of one or two packages, depending on the model and manufacturer, with mounting brackets designed for easy alignment of the sensors. Figure 5–13 shows the internal arrangement of parts and components of an Omron model E3B sensor. The packages are designed for operation under

Automation Sensors **169**

AMPLIFIER SELF-CONTAINED TYPE			AMPLIFIER SEPARATED TYPE	Color mark sensor
E3S-G	E3S-L	E3S-X	OPE/ORE	E3ML
Grooved head type ideal for mark sensing and positioning	Focusable type with built-in amplifier	Fiber optics type for high resolution sensing	Subminiature series ideal for OEM use	High-performance color mark sensor
E3S-GM5E□ E3S-GS3E□			OPE-S100	
Grooved head type	Definite reflection type	Separate type Diffuse reflection type	Separate type Diffuse reflection type	Separate type Diffuse reflection type
12 VDC −10% to 24 VDC +10%	12 VDC −10% to 24 VDC +10%	12 VDC −10% to 24 VDC +10%	120/240 VAC ±10%, 50/60Hz (with amplifier unit)	10 to 30 VDC (For lamp 4.5 VAC)
40mA max.	50mA max.	50mA max.	Approx. 5VA	40mA max. (at 30 VDC)
5mm, 3cm	3 to 10cm or 5 to 25cm	8mm, 3cm	3, 5 or 10cm 1 or 3m	0.5 to 20mm
Marks on transparent sheet Opaque and translucent materials	Transparent and opaque materials	Transparent, translucent and opaque materials	Transparent, translucent and opaque materials	Any color mark
Solid-state output: Output (source) current: 1.5 to 3mA Load (sink) current: 80mA max.	Solid-state output: Output (source) current: 1.5 to 3mA Load (sink) current: 80mA max.	Solid-state output: Output (source) current: 1.5 to 3mA Load (sink) current: 80mA max.	Contact output: 230 VAC 1A Solid-state output: 15 VDC Output impedance: 3.3kΩ	Solid-state output: Load current: 80mA max. Output impedance: 4.7kΩ
2msec max.	5msec max.	2msec max.	Contact output: 30msec max. Solid-state output: 3msec max.	20μsec max.
Operating: −25 to +55°C	Operating: −25 to +55°C	Operating: −25 to +55°C	Operating: −25 to +55°C	Operating: −10 to 55°C
Water-resistant type	Water-resistant type	Water-resistant type	Dust-proof type	Immersion-proof type
IP66	IP66	IP66	IP51	IP67
Types 1, 4, 4X, 12	Types 1, 4, 4X, 12	Types 1, 4, 4X, 12	—	Types 1, 4, 4X, 12
(UL) —	(UL) ●	(UL) —	(UL) ●	—
351	355	361	365	371

Figure 5–12 *(Continued)*

varying environmental conditions, and models can be found that meet world standards and those of the National Electrical Manufacturers Association (NEMA). The sensors include solid-state sensing electronics, which drive the light source in the transmitter, and a light-sensitive element in the receiver. The electronics require dc or ac voltage for power (typically 12 to 24 volts regulated dc or 90 to 250 volts ac). The sensor output includes an electrical signal and light-emitting diode to indicate the state of the sensor. The electrical signal can drive a sensor controller that provides relay contacts for use in switching higher current loads or can drive smaller loads directly. The Omron models, for example, have direct output drive capability that ranges from 80 milliamperes on some models to 200 milliamperes on the larger units. The response time of photoelectric sensors is in the 2- to 30-millisecond range.

Figure 5–14 shows several examples of sensors being used to check for the presence of parts. These could very easily be incorporated into a robot work cell. For example, Figure 5–14A, C, G, J, and K illustrate how a sensor could be used to

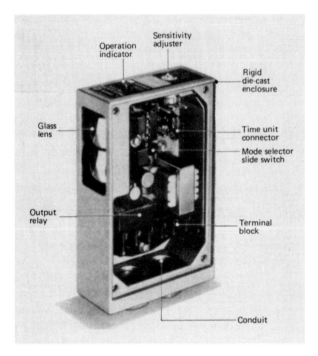

Operation indicator

Sensitivity adjuster

Rigid die-cast enclosure

Glass lens

Time unit connector

Mode selector slide switch

Output relay

Terminal block

Conduit

Figure 5–13 Sensor Construction. (Courtesy of Omron Electronics Inc.)

check for the presence of various types of parts a robot might be moving in a material-handling application or using in an assembly application. The application in Figure 5–14B involves the tasks of unstacking thin plates. The signal from the sensor is used by the robot controller to stop the downward motion of the arm and activate a vacuum gripper. In a robot drilling application, the drill length must be checked frequently to be sure a drill rod has not been broken. The application in Figure 5–14D shows how this is accomplished with a photoelectric sensor.

Application of photoelectric sensors involves the same four steps described at the end of the section on proximity sensors. The factors affecting sensor selection include the following seven:

1. Sensing distance required by the application
2. Sensor mounting requirement
3. Work-cell area available
4. Size, shape, and surface reflectivity of the part to be sensed
5. Response time required
6. Environmental conditions in the work cell, especially background light present
7. Interface requirements

Selecting the sensor that best satisfies the work-cell conditions requires a thorough study of the application. The primary elements of the study are the factors just listed, and the results of the study are criteria that can be used to select the sensor model for the job.

- When the sensor is susceptible to the reflection from background object surface

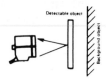

Typical examples
(1) Sensing of thin objects on the conveyor line.
(2) Sensing of objects in the presence of a background object with high reflection factor such as rollers, metallic plates, etc.
(3) Sensing of the residual quantity in a hopper or a parts feeder.

(A)

- Sensing of level or height

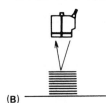

Typical examples
(1) Sensing the height of stacked plywood, tiles, etc. from above.
(2) Monitoring and control of the liquid level from above.
(3) Determination of the heights of objects on a conveyer line.
(4) Sensing of slack in sheets from above.

(B)

- Sensing of objects traveling in contiguous succession

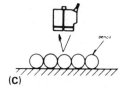

Typical examples
(1) One-by-one sensing of pencils or metallic bars traveling successively or in contiguous succession.
(2) Similarly, one-by-one lateral sensing of bottles or cans traveling in contiguous succession.

(C)

- Sensing of small, slender or fine objects

Typical examples
(1) Sensing of broken drill bits.
(2) Sensing of small parts such as electronic components.
(3) Sensing of the presence or absence of bottlecaps.
(4) Sensing of fine mesh.

(D)

- Sensing of small holes, narrow openings, or unevenness

Typical examples
(1) Sensing of holes in flat board.
(2) Sensing of protrusions.

(E)

- Sensing of objects utilizing their difference in luster

Typical examples
(1) Identifying the face or back of tiles.
(2) Identifying the face or back of lids.

(F)

- Sensing of transparent objects

Typical examples
(1) Sensing of transparent or translucent objects.
(2) Sensing of transparent glasses, film, or plastic plates.
(3) Sensing of the liquid level.

(G)

- Sensing of objects through a transparent cover

Typical examples
(1) Sensing of the contents in a transparent case.
(2) Sensing of the position of meter pointer.

(H)

- Sensing of the edge of object

Typical examples
(1) Positioning control of plywood
(2) Positioning control of various other products.

(I)

- **For sensing of presence of parts in parts feeder**

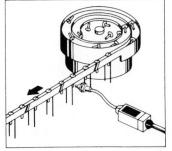

(J)

- **For sensing of presence of resistors on conveyer line**

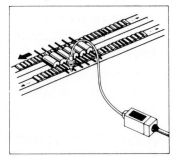

(K)

Figure 5–14 Example Applications of Photoelectric Sensors. (Courtesy of Omron Electronics Inc.)

The sensitivity of photoelectric sensors can be adjusted so that the sensing of transparent, translucent, and opaque materials is possible with the same model sensor. The procedure for adjustment of sensor sensitivity varies according to models and manufacturers. On some models the indicator LED (light-emitting diode) changes colors as the sensitivity is adjusted; on others, the LED changes from *on* to *off* to indicate the sensing condition. On all models with a sensitivity adjustment, the sensor can be operated in a stable mode with a variety of surface finishes on the part being sensed.

Interfacing the sensor to supply power and controller inputs requires the designer first to consider the sensor model selected. If regulated dc between 12 and 24 volts is available, then a sensor controller is not required. Without a sensor controller the sensor output must be interfaced directly to robot controller input or the input of a programmable logic controller. This requires an output circuit in the sensor that is compatible to the input circuit in the robot or programmable logic controller.

5-4 PROCESS SENSORS

In addition to the sensors required for the robot, most process or manufacturing operations need sensors to monitor parameters inherent in the process itself. Also, many of the other automated machines have sensors to alert and warn the operator about conditions that are developing within the operation. A forging operation is a good example. Through warning lights, the human operator monitors the temperature of the oven that is heating the parts before forging. The level of oil and oil pressure for the press are also displayed by indicators. With a robot present, the visual indicators are of no value. If visual indicators are replaced with electronic sensors, however, then an electrical signal can be used to alert the robot or work-cell controller that a situation exists that needs corrective action.

The only requirement for the system or machine parameter to be monitored by the robot is that it must be discrete, that is, either an *on* or *off* signal. Analog sensors that have a variable output, 0 to 5 volts, for example, are associated with proportional control requirements. Although these could be interfaced to the work cell or robot controller, they are usually handled by a separate control system designed to provide proportional control. A discrete signal, called an *alarm*, could be provided by the proportional controller to warn the robot that the process is off course and a correction is necessary.

CASE: AUTOMATION AT WEST-ELECTRIC

Marci looked at her watch. With just 30 minutes before the meeting with Bill, she had better review her suggestions for the performance measures and the technical issues associated with the automation of the upset forging area. She took a folder out of her desk drawer and began to review the data.

PERFORMANCE MEASURES

The performance measures for the upset forging area with current values:

Lot size	1500 parts
Average die life	1754 machine cycles
Productivity (average)	1105 parts per shift per work cell

The lot size is critical because it dictates the inventory levels that must be carried and the shipment sizes that must be imposed on our internal customers. The lot size is set by setup time and production cost, and so a reduction in setup time would permit smaller lot sizes. Quality is a factor in the average die life performance measure, because part variation becomes excessive when die wear exceeds specified limits. Frequently, dies are used beyond the optimum limit to finish a production run at the current lot size. The 70,000 parts per week target production goal requires 1555 parts per work cell per shift. The current production average, 1105, is 450 parts short of the goal. Overtime is required to make up the difference when demand is high. In addition, the manual production rate is not consistent because the initial rate at the start of the shift is 1400 parts per shift (17-second cycle time), and at the end of the shift it is 1054 parts (23-second cycle time). ●

TECHNICAL ISSUES

Performance Requirements

- *Cycle times:* Cycle times must be 13 seconds throughout the shift with zero time lost for die change and machine setup to meet the required production level.
- *Part-handling specifications:* The end of the extrusion with the large diameter is the optimum location for gripping by the robot for movement in the cell.
- *Human backup requirements:* In the event of a robot failure, the cell must be capable of limited production using human operators.
- *Future production requirements:* Two future production requirements must be considered: (1) the reduction of cycle time to 10 seconds, and (2) the integration of the extrusion and upset forging operations.
- *Malfunction routines:* The control software must include a three-light system: the first light indicates low raw material stock, the second indicates an operational problem (part jammed, oven temperature low, poor quality, etc), and the third indicates production equipment failure.
- *Allowable downtime:* A 13-second cycle permits about 6 hours per week for preventive maintenance and unplanned downtime.
- *Maximum repair time:* A goal of 2 hours to repair or replace any of the hardware in the cell is needed.

Layout Requirements

- *Service and accessibility for maintenance:* Whenever possible, the control hardware for the cell should be located outside the working envelope of the robot.
- *Equipment relocation requirements:* If the new cell area is prepared, with the robot, parts feeder, and inspection system in place, then the relocation of the forges and ovens can be completed in 2 days. An additional 2 days are required to fine-tune the system with only limited production from the automated cell.
- *Floor loading:* The floor can support the new cell design.

Production Characteristics

- *Inspection needs:* The testing of one part per hundred must be supported by the automation. The use of smart gages and an automated fixture is recommended.

Equipment Modifications

- *Requirements for unattended operation:* The system must operate three shifts with a resupply of the raw material every $2\frac{1}{2}$ hours. Changes in production models and stocking of the oven must be supported by the work-cell software.
- *Requirements for increased productivity:* The number of positions in the oven and system components must be provided for a 20 percent increase in productivity in the future.

Process Modifications

- *Lot-size changes:* A reduction of lot size from 1500 to 1000 is required on the initial design.
- *Cell operation:* The robots, ovens, parts feeders, inspection devices, and two of the three forges will operate continuously over three shifts. One forge will always be idle to permit die change and setup to minimize production downtime. The proposed operation for the cell in Figure 4–33 is described by the schedule in Figure 5–15. Note that robots *A* and *B* use forge *C* as an alternate work area after completing production on forges *A* and *B*. When the robots move to forge *C*, the dies are changed on forges *A* and *B*. At one point in the schedule, forge *C* is not in use, which allows the dies to be serviced in that forge for the next scheduled part.

As Bill hung up the phone he looked up to see Marci standing at his office door. "Marci, come in and pull up one of those chairs. I was just talking to Mike; we are going to work on the robot specs later this afternoon. I'm anxious to see what you have on our technical requirements."

As Marci handed Bill a folder, she said, "The work is not complete, but there should be enough there for you and Mike to use. Look it over to see if you have any questions."

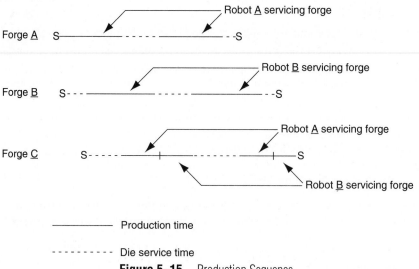

Production time ——————————

Die service time - - - - - - - - -

Figure 5–15 Production Sequence.

After studying the two documents, Bill said, "I'm surprised that you didn't recommend a quality target in the performance measures. What's your rationale for the ones you listed?"

"Bill, I want to address quality by reducing lot sizes and increasing die life. Let me explain. We measure a part every hundred cycles to determine when the die is showing excessive wear. However, we often finish out the production lot with a marginal die, and produce one or two hundred parts out of tolerance." Marci paused to let her comment sink in, then said, "Manufacturing argues that at this stage in production it doesn't make any difference. The truth is we really don't know. I want to work on consistent cycle times that should improve die wear and get the lot sizes down through a reduction in setup time. Then we can be sure we run only on good dies, and see if downstream quality improves. I'm betting that it will," she said.

Bill felt that Roger was looking for a quality statement in the performance measures, but after discussing the alternatives, he agreed with Marci's analysis. He flipped to the technical issue list and said, "I don't see a path velocity for the part to complete the 15-second cycle time. Have you got a number in mind?"

"I would rather wait for Ted's simulation before we settle on a value. It should be ready tomorrow," Marci said. "My estimate is a path length of about 15 feet for the primary forge and 25 feet for the shared one. So . . . 2 feet per second should do the job." After clarifying several other values on the technical issues list, Marci gathered her material and left.

After the meeting with Mike to define a general set of robot characteristics, Bill stopped by Jerry's office and picked up the proposed modifications for the oven. As soon as he received the work-cell layout from the simulation, he would have the data necessary to put an initial design together. He spent the rest of the day preparing copies of the data for the team so that they could review each other's work before their next meeting. In addition, he modified a sensor selection checklist, shown in Table 5–1, from a technical journal and asked the team to compile a list of sensors that would be needed in the cell. The following material was sent to the team.

Table 5–1 Sensor Selection Checklist

Target information

Parameter	Data needed
Material	Is the target material ferrous, nonferrous, or nonmetallic?
Mass	How much does it weigh?
Size	How large is the target in square inches or cubic inches?
Shape	Is the target flat, cylindrical, cubic, or spherical? Is it compact, long, or narrow?
Surface	Is the surface opaque or transparent? If opaque, is it colored, shiny, dull, smooth, textured, porous, or nonporous?
Motion	What is the velocity in feet per second? Is the direction of movement perpendicular or oblique to the line of sight? If using photoelectric sensors, can the reflector or receiver be mounted behind the target?
Distance	How far away is the object to be detected? How close to the part or machine can the sensor be located? (This parameter alone can determine whether a proximity or photoelectric sensor should be used.)
Presentation	What is the rate of presentation to the sensor in units per minute? Are targets oriented randomly or regularly? Are parts separated or do they overlap? What is in back of the target?
Precision	Is precise range needed? This is usually true in machine tool applications.

Environmental information

Parameter	Data needed
Enclosure	Is a NEMA qualified case required? Is the material aluminum or plastic?
Temperature	What is the typical annual range in °F? Is it very cold or very hot?
Relative humidity	What is the typical annual range in percent? Is it very dry or very wet?
Ambient lighting	What are the sources of light? Are they constant or variable? Is there any direct sunlight?
Electrical noise	What are the typical and peak intensities? A welding line would represent a "worst-case" example.
Mechanical vibration	What is the typical amplitude range? What is the maximum physical shock possible?
Air quality	Is the air clean or dirty? Is the contaminant dust, smoke, oil mist, or paint spray? Is the atmosphere explosive?
Mounting surface	Is the material ferrous, nonferrous, or nonmetallic? How thick is it? Is it flat, rounded, or some other shape? How much mounting space is available in square or cubic inches? Are precision clearances necessary? Is the mounting surface stationary or mobile?

Electrical circuit

Parameter	Data needed
Power supply	120 VAC, 50/60 Hz, or 12/24/35 VDC?
Contacts	Number and configuration?
Output form	Analog or digital?
Output load	Minimum and maximum current required?
Output voltage	Minimum and maximum voltage required?
Switching delay	Activation delay in microseconds? Release delay in microseconds?
Switching frequency	Hz?

(Reprinted with permission from *Sensors—The Journal of Machine Perception,* April 1986.)

Robot Specifications

Position repeatability is a critical parameter in this application, and a robot with a repeatability of 0.005 satisfies work-cell requirements. A jointed spherical arm with 6 degrees of freedom is necessary to reach the numerous taught positions at various elevations and orientations. The work envelope requires a maximum reach of 5 to 6 feet from the robot's center axis and a rated payload of 20 to 25 pounds. The maximum velocity must be 4 feet per second or greater so that a 50 percent safety factor can be applied. Both on-and off-line programming are required, along with a controller that supports easy interfacing to a programmable logic controller. The cost must be in the $40,000 to $60,000 range to achieve the desired payback. Compliance, force/torque sensing, vision, specified downward force, and tool changing are not considered necessary for the application.

OVEN MODIFICATIONS

The ovens used throughout the turbine blade production line have 30-inch steel tables with a center axle that rides in a bearing above and below the heated area of the oven. A motor, sprocket, and chain drive system, located in the base of the oven, indexes the table every time the door is opened and closed. The ovens would be modified to support automated operation by using a slotted table to hold the extruded parts during heating. The hanging orientation supports loading and unloading of the oven with robot automation. The suggested configuration for the

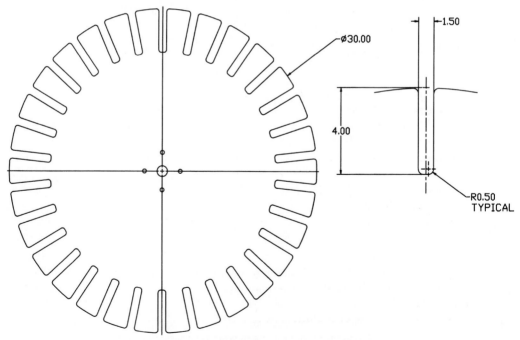

Figure 5–16   Oven Part Holder for Automated Operation.

table, Figure 5–16, has 30 slots; however, additional slots could be added for higher productivity. The dc motor, currently used to turn the table, would be replaced with a stepper motor for accurate and repeatable rotation. The oven door foot switch would be interfaced to a programmable logic controller, and the oven temperature gage would be replaced with a sensor for automatic monitoring of oven temperature. Warping of the slotted table during repeated heating and cooling of the oven is the major design problem to solve. Engineering is researching the best material to use for production of the table to minimize the warping problem.

Work-Cell Simulation

The work-cell simulation illustrated in Figure 5–17 will work with an ABB Robotics model IRB 2400 robot. Other brands and models are available, but the specifications of the model 2400 closely match or exceed the initial robot specifications for the work cell. After the parts feeder and inspection station are defined, the following criteria will be established: programming of the operation, generation of minimum cycle times, optimum position of hardware, and axes movement limits during operation with each forge.

5-6 SUMMARY

Sensors make the robot a part of the environment in which it exists. They give the robot information about the work cell that is vital to normal operation. Sensors are used to (1) protect worker and robot from harm, (2) monitor the production system and work-cell operation, (3) analyze product quality, and (4) provide part identification and orientation. All sensors are grouped into either a contact or noncontact category, and each type can have either a discrete or an analog output signal. Contact sensors include limit switches and artificial skin, and the most commonly used noncontact sensors include proximity sensors and photoelectric devices. Limit switches are available with tripping mechanisms, including plungers and levers. Contacts are rated as either pilot duty or electronic duty, based on the level of current that is controlled. A large selection of devices supports the many different operational characteristics present in automation. The concept of artificial skin focuses on the need for a sensor that can replicate the sense of touch used by human operators. Sensors in this area provide either simple touch sensing or full tactile sensing; the latter is difficult to achieve at this time.

Proximity sensors detect the presence of a part when the part comes within a specified range of the device. The sensors are available in a variety of package shapes and sizes. The sensing technology uses either a magnetic field (metallic part) or electrostatic capacitance (both metallic and nonmetallic parts) to detect the presence of a part. Output circuits dc and ac, in normally closed and normally open configurations are available. The dc output in either an NPN or a PNP transistor configuration is used for smaller power requirements, and the ac output is used for larger loads.

Photoelectric sensors detect the presence of a part when a transmitted light beam is broken or reflected to a receiver. The four types commonly used include

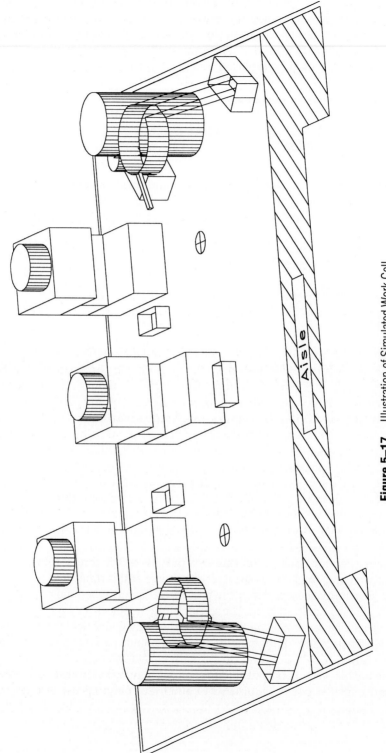

Figure 5–17 Illustration of Simulated Work Cell.

separate, retroreflective, diffuse reflective, and definite reflective. The operational characteristics for the four types distinguish each category; however, the primary difference is range or distance to the detected part. As is true of proximity sensors, a variety of package shapes and output configurations are available.

Process sensors measure parameters that are part of the production process. In most applications the sensors have analog output signals to provide the exact value of the measured parameter.

QUESTIONS

1. What is the basic function of sensors in an automated work cell both with and without a robot present?
2. What are the six basic reasons sensors are used in a work cell?
3. What are the two categories into which all sensors can be grouped?
4. What kinds of sensors are included in the contact sensor group?
5. What is the definition of a limit switch?
6. What are the four basic parts of every limit switch?
7. How does pilot duty differ from electronic duty in the electrical characteristics of limit switches?
8. Using a data sheet for a general-purpose limit switch, determine the switch part number and lever part number required for rotary operation using fork rollers on the same side, and double-pole–double-throw operation for electronic duty.
9. What is the function of the lamp on a limit switch?
10. What is the distinction between tactile sensing and simple touch sensing?
11. Describe how tactile sensors operate.
12. What three steps are required for intelligent acquisition of parts?
13. What sensors are included in the noncontact sensor group?
14. Describe the four package configurations now available with proximity sensors.
15. Describe the difference between metallic detection and nonmetallic detection proximity sensors.
16. What three output circuits are available on proximity sensors?
17. Describe the four different types of photoelectric sensors.
18. What is the primary difference among the four types of photoelectric sensors?
19. What type of photoelectric sensor has the greatest range?
20. What type of photoelectric sensor is most accurate when used for range finding?
21. What factors affect photoelectric sensor selection?
22. Select a photoelectric sensor and sensor controller for the following applications. Identify the part numbers of the sensor and controller required.
    (a) A sensor to operate at a distance of 8 centimeters with high background light. The output should be *on* when the object is sensed.

(b) A sensor to operate over a distance of 8 feet to detect when a part breaks the beam. The output should be active when the beam is broken.

(c) A sensor to count the number of resistors in a parts feeder by using the resistor leads to reflect a beam. The output should be *off* when the lead is present.

(d) A sensor to count revolutions of a shaft using marks on a transparent disk mounted to the shaft. The output should be active when the mark is not present.

PROBLEMS

1. Draw a classification tree that includes all the sensors described in this chapter.
2. Design a decision tree to select the best sensor for a given manufacturing problem. Use questions at each branch of the tree that require either a *yes* or *no* response.
3. Write a computer program that will execute the decision tree from problem 2.

CASE PROJECTS

1. Using the robot checklist data developed by Bill and Mike in the case, identify two robots that could be used in the upset forging work-cell.
2. Using the robot checklist data developed in question 8 of the case/projects section of Chapter 4, identify two robots that could be used in the slug lubrication and extrusion work-cell.
3. Using case data, the sensor selection guidelines, and the cell design for upset forging developed by the W-E team, complete a table that has the following information: sensor number, work-cell condition to be sensed, and the type and style of sensor (limit switch, lever, proximity, and photoelectric) to be used. Then comment regarding the operation in the cell.
4. Using case data, the sensor selection guidelines, the six basic reasons for sensors in an automated cell, and your work cell developed in question 2 of the case/projects section in Chapter 4, complete a table that has the following information: sensor number, work-cell condition to be sensed, the type and style of sensor (limit switch, lever, proximity, and photoelectric) to be used. Then comment regarding the operation in the cell.
5. Describe any differences between the robot checklist data developed by Bill and Mike in the case study and your answer from question 7 of the case/projects section in Chapter 4.
6. The team is looking strongly at the ABB IRB 2400 robot. What other robot models should be considered that would satisfy the criteria in the case study?
7. How well will Marci's performance measures for the upset forging cell indicate that the automation has improved the blade manufacturing performance? Should other performance measures be included?

8. Bill accepted Marci's recommendations for the performance measures on the upset forging cell. Should he have insisted on a performance measure directly related to quality? Why or why not?

9. Using the definition for robot payload from Chapter 1, the weight of the gripper fingers calculated in question 4 of the case/problems section in Chapter 4, the maximum part weight, and the value Bill and Mike specified for robot payload range, determine the allowable weight for the parallel gripper. Will Ted's finger designs work? What are some other options?

Work-Cell Support Systems

6-1 INTRODUCTION

Some combination of the sensors described in the previous chapter are found in work cells that use robot automation. In addition to basic contact and noncontact sensors, many of the cells have a variety of other support systems present. A description of all the supporting equipment used in robot cells is beyond the scope of this text, but we will discuss some of the systems that are used frequently. The systems associated with robot automation include *vision, material handling, part feeding, inspection, automatic tracking,* and *safety.* The first five support system areas are covered in this chapter, and the sixth, safety, is included in Chapter 10.

6-2 MACHINE VISION SYSTEMS

Vision systems are being used increasingly with robot automation to perform the following tasks:

- *Part identification:* Commercially available vision systems store data for different parts in active memory and use the data to distinguish between parts as they enter the work cell. The system can learn the characteristics of different parts and identify each part from its two-dimensional silhouette.
- *Part location:* Vision technology allows the user to locate randomly placed parts on an X-Y grid. The vision system measures the X and Y distances

from the center of the camera coordinate system to the center of the randomly placed part.

■ *Part orientation:* Every part must be gripped in a specified manner by the end-of-arm tooling. The vision system supplies the orientation information and data that are used to drive the gripper into the correct orientation for part pickup. Many part orientation parameters, both measured and calculated, are provided by the vision system for use in automated part handling.

■ *Part inspection:* Vision systems are used to check parts for dimensional accuracy and geometrical integrity. The parts are measured by the camera, and the dimensions are calculated; at the same time, the vision system checks the parts for any missing holes or changes in the part geometry.

■ *Range finding:* In some applications the system uses two or more cameras to measure the X, Y, Z location of the part. This technique is also used to measure and calculate the cross-sectional area of parts.

The use of vision to enhance the operation of an automated work cell has moved from the research laboratory to the factory floor. More than fifteen manufacturers provide equipment that gives robots the eyes they need to perform complex manufacturing tasks. Most of the equipment is for part identification, location, and orientation information necessary for automatic handling of parts plus inspection.

Vision Standards

The standards established by the AIA for the RIA include the following six ANSI standards:

1. AIA A15.08/1: Standard AIA Analog Camera Connectors.
2. AIA A15.08/2: Standard AIA Digital Camera Connectors.
3. AIA A15.08/3: Monochrome Digital Interface Specification.
4. AIA A15.08/4: RGB Digital Interface Specification.
5. AIA A15.05/5: Monochrome Analog Interface Specification.
6. AIA A15.05/6: RGB Analog Interface Specification.

Vision System Components

The block diagrams in Figure 6–1 through 6–3 illustrate three architectures used to implement vision technology. In Figure 6–1, a self-contained system is illustrated. In this type, the vision system does not use resources from any other work-cell hardware. Note that vision information is passed to the work-cell robot through a serial interface. The system in Figure 6–2 has the vision system integrated into the work-cell programmable logic controller (PLC), in this case an Allen Bradley PLC-5. This type of system has the vision data and PLC control data on the same electrical information bus, called the *back plane*, inside the PLC. The configuration in Figure 6–3 integrates the vision system with the robot controller,

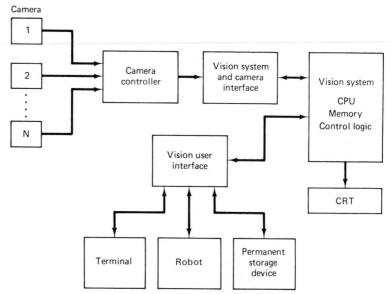

Figure 6–1   Block Diagram for Stand-Alone Vision System.

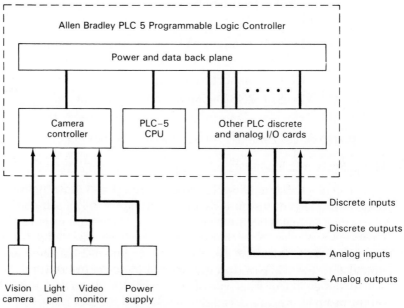

Figure 6–2   Machine Vision System with Programmable Logic Controller (PLC).

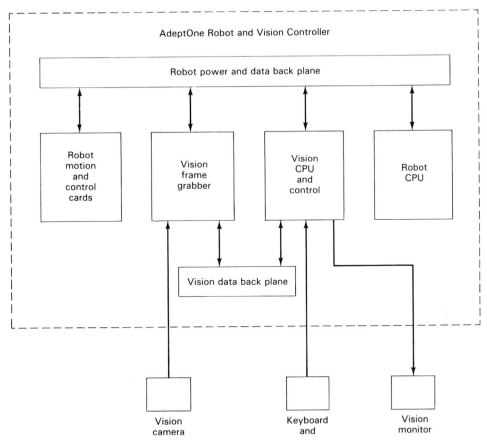

Figure 6–3 Adept Integrated Robot and Vision System.

a technique used by Adept Technology, Inc. for its robot vision systems. Note that the vision system has a separate back plane for exchanging vision data between the electronic cards in the vision system.

The singular advantage of the Adept integrated vision system is the significant improvement in response time for the host system when it responds to changes in the work cell sensed by the vision system. A secondary advantage is the reduced cost from fewer interfaces and less complex electronics required when integration is present.

The basic vision system components illustrated in all the configurations include the following: one or more cameras, a camera controller, interface circuits and systems for the camera and work-cell equipment, a high-resolution cathode ray tube (CRT) for image display, and a lighting system for the parts.

The lighting system illuminates the critical features of the part so that they can be captured by the vision system cameras. The vision camera and lens system measure the level of light coming from the part under study; they capture a black and white representation of the part. A charge-coupled device (CCD) camera is used most frequently, but in some cases a vidicon tube camera is used. These two types of cameras are very different in operation.

The vidicon tube used in vision applications produces an analog output signal much like the cameras used in the broadcast television industry. The vidicon is a vacuum tube with a flat circular end designed to receive the image of the part from the lens. An internal electron beam is scanned across the image of the part at the end of the tube, and the level of light in the image produces a proportional analog voltage for every point that the beam strikes. If large image areas are important, then the vidicon is the most cost-effective solution. However, such a camera is not as linear or as stable as a CCD device and can have a 10 percent error due to image distortion.

The CCD camera uses a solid-state array of light-sensitive cells deposited on an integrated circuit substrate. Each cell is a small light-sensitive device whose output is a function of the intensity of light striking its surface. The CCD systems come in two basic configurations: *linear* arrays and *imaging* arrays. Figure 6–4 shows how each of the types is organized. The linear arrays measure a single line, whereas the imaging arrays measure a complete two-dimensional image and cost more than the linear type. CCDs are accurate, rugged, and have good linearity.

Image Measurement

The basic unit of measurement in a vision system is the *gray scale;* the basic parameter measured is *light intensity;* and the basic measurement element is the *pixel.* The vision system lens focuses light from the part onto the light-sensitive surface in the camera. The light-sensitive surface is divided up into small regions or picture cells. Each of these regions or picture cells is called a *pixel.* The resolution of the vision system is directly proportional to the number of pixels on the light-sensitive surface. For example, a sensor with a 256 × 256 array will have a higher resolution than a sensor with a 128 × 128 array of pixels. Higher resolution means greater accuracy when a system is used to measure part dimensions. Measurement is best understood with an example. If a 1-inch square produces an image that fills

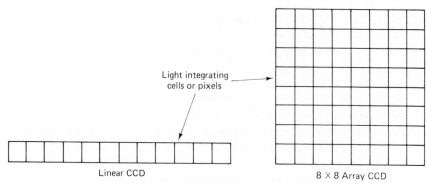

Figure 6–4   Charge-Coupled Devices (CCD).

the 8×8 CCD array in Figure 6–4, then the smallest variation that could be detected is $\frac{1}{8}$ inch. If a 32×32 array with the same dimensions was substituted for the array in Figure 6–4, then the resolution of the measurement would be $\pm 1/32$ inch. Each pixel on the sensor's surface is excited by the light focused on it by the camera lens. With no light present the pixel is turned *off*, but when light reaches a *saturation level* the pixel is full *on*. Between those two extremes there are shades of gray that cause the pixel to be excited to a partially *on* condition. The number of excitation states between *off* and *on* is called the gray scale. Current systems have gray scales ranging from 4 to 256. The greater the number in the gray scale, the better the system is at locating fuzzy edges and shadows of parts. The gray scale of a pixel is the numerical representation of brightness for one small spot on the part.

Image Analysis

The first step in image analysis is to locate all the areas in the image that correspond to the part being viewed by the camera. Most current vision systems are two-dimensional, which means that the outline of the object is the only feature of interest to the system. The two-dimensional features of a part are enhanced by placing the part on a backlighted surface or by placing the object on a surface with contrasting color and using top lighting. Figure 6–5a shows a part on a backlighted surface, and Figure 6–5b shows the corresponding image generated by the vision system. The two most commonly used systems to find objects from camera data are *edge detection* and *clustering*. Both techniques attempt to locate the boundaries of objects or regions in the image so that their location, size, shape, and orientation can be computed as clues for recognition. Edge detection is based on the fact that there is a sharp difference in brightness between the object and its background. This can be seen in Figure 6–5a. Areas of high contrast are found by searching through the array of pixel data for jumps in the gray level. To locate the change in gray level, a mathematical operator examines a small neighborhood of adjacent pixels and computes an *edge probability* value. If the value is above a set threshold,

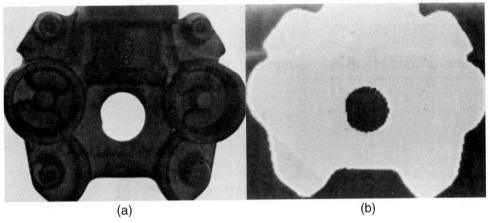

(a)   (b)

Figure 6–5   Vision System Representation of Production Casting.

then a notation is stored in the vision controller memory and part data indicating a probable part edge. The ability of the system to eliminate false edges because of electrical noise and to find fuzzy edges is improved if the size of the neighborhood used to compute the edge probability has many pixels. Because increasing the size of the neighborhood increases the computation time for the system, a trade-off is required.

The second method used to find objects from an array of pixel data is called *clustering* or *region growing*. The system basically tries to find adjacent pixels that have similar properties. This technique uses a *discrimination function* to determine if a pixel is part of a region being developed. The discrimination function computes the desired properties of a pixel and compares them with the properties of other pixels in the region or against a threshold value. In the simplest case, the gray level of each pixel could be checked against a threshold value to determine if the pixel is in the part image or in the background area. Many vision systems locate parts by using every pixel whose gray level value is different from the gray level value for the background. This process starts with a *seed* pixel within the part image and then grows in every direction until a boundary is reached. After the area of the part has been developed, the pixel data are frequently changed to binary values; that is, every pixel that is part of the object is stored as a 1, and every pixel that is not part of the object is stored as a 0. Transferring these data to a CRT produces the part silhouette shown in Figure 6–5b.

Image Recognition

The second problem facing the vision system after the edges are detected and stored is the *recognition* or *identification* of the current image. Three frequently used two-dimensional recognition strategies are template matching, edge and region statistics, and statistical matching using the Stanford Research Institute (SRI) algorithm or one of the many others developed in the past 15 years.

In the *template matching* technique, templates of the parts to be recognized are stored in the vision system memory. The images recorded by the vision camera are compared with the templates stored in memory to determine if a matching part is present. Figure 6–6 shows the template of a part and the problem that this type of system must overcome. Any change in the object's scale or a rotation of the view makes a match with the template more difficult. The Allen Bradley vision system

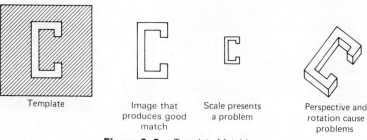

Template   Image that produces good match   Scale presents a problem   Perspective and rotation cause problems

Figure 6–6   Template Matching.

illustrated in Figure 6–2 uses the template matching process. In the teaching process, regions in the captured image are identified and recorded with the correct contents or parts present. In operation, an image of the region is captured and compared with the image recorded during the teaching process. If the image values for the regions are the same, the captured image matches the stored template, and the part is identified.

The *edge and region statistics* technique defines significant features for the parts studied, then develops a method of evaluating these features from the pixel data. Some of the more common features used in this method are the following:

- *Center of area:* A unique point from which all other points on the object are referenced
- *Major axis:* The major axis of an equivalent or best-fit ellipse
- *Minor axis:* The minor axis of an equivalent or best-fit ellipse
- *Number of holes:* The number of holes in the object's interior
- *Angular relationships:* The angular relationship of all major features to one another
- *Perimeter squared divided by the area:* A value unique to any given shape that does not change as the image of the object changes scale
- *Surface texture:* An identifying characteristic for which gray level images are used

The statistical data for parts to be identified are stored in memory, and when the camera produces an image of an unknown part the system calculates a set of feature values from the pixel data. The feature values for the unknown part are compared statistically with the feature values stored in memory. If a close match is found, then the part is recognized and identified by the system. The SRI technique is similar to the method just described. The one exception is that an algorithm is used to apply the features identified for each part.

Vision will be an important tool in future automated work cells, but a significant amount of work remains to be done before it can perform the basic operations now done by human operators. The areas requiring additional work include three-dimensional camera systems, ranging techniques, tracking, and handling missing or extra features in taught objects.

6-3 LIGHTING FOR MACHINE VISION

Although lighting is an important consideration in every vision application, it is frequently overlooked. In many cases a greater research effort on the lighting and optics problem would result in a less sophisticated and lower cost vision system.

Selection of the Light Source

The light source must provide the vision system with the best possible images of the part under production conditions. The analysis of captured images requires high contrast between part features and the background of the captured data. To achieve the high contrast, lighting systems must minimize the effects of natural light, shop lighting, and radiation from process sources that could disrupt the vision system operation. Selection of a lighting source for a vision application is driven by three factors: (1) the type of features that must be captured by the vision system; (2) the need for the part to be either moving or stationary when the image is made; and (3) the degree of visibility of the environment in which the image must be captured. The three lighting techniques used to satisfy these factors are described next.

Lighting Techniques

The three lighting techniques used in vision applications are *front lighting, backlighting, and structured lighting.* In front lighting, Figure 6–7a, the camera and light source are on the same side of the part. This technique, the most common in vision applications, is used when the surface features of the part are important image characteristics. (Review Figure 3–6 and the associated case study and note the front lighting used.) Continuous front lighting is used when the part is stationary or is moving slowly, whereas strobing is used to capture images of parts moving rapidly past the camera. The strobe system bathes the part in a high-intensity flash to freeze the fast-moving part and capture an image with a synchronized video camera.

The backlighting technique, Figure 6–7b, provides an image with the greatest contrast; however, only the silhouette of the part is captured by the vision system. The part in Figure 6–5 is backlighted, and the silhouette is clearly visible. The high contrast provided by backlighting makes the identification of part edges less complex. This lighting technique is limited to vision inspection applications for measuring a silhouetted feature or determining the presence or absence of a feature. Again, the backlight source is continuous if the parts are stationary, and a strobe is used if the parts are moving when the image is captured.

The third technique, structured light, is illustrated in Figure 6 –7c. Similar to front lighting, in structured light the light source and camera are on the same side of the part. However, a structured light source controls the shape and form of the projected beam to aid in the capture of part features. The light source in Figure 6–7c, for example, produces a narrow slit of light which is projected onto the part at an angle relative to the camera. Light beams are controlled with apertures, lenses, fiber optics, and coherent light sources such as lasers. Structured light is used in two application areas: (1) the capture of a specific part feature from a complex shape when only the feature of interest is lighted, and (2) the extraction of three-dimensional information from a part using only a two-dimension vision system. The second application is illustrated in Figure 6–7c. Note that the angular shape on the surface of the part is visible to the camera because the narrow band of light is shifted due to the angle of the light source.

Illumination Sources

The illumination energy source chosen for the vision application is equally important for a successful project. The major types of sources include these:

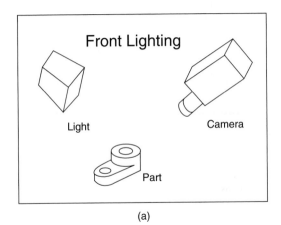

(a)

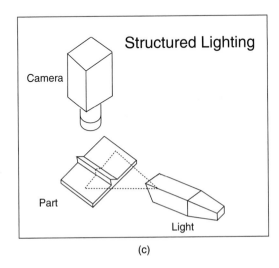

(b)

(c)

Figure 6–7 Lighting for Machine Vision.

incandescent bulbs, fluorescent tubes, xenon flash tubes, lasers, infrared light-emitting diodes, and X-ray tubes.

Each of the sources has unique properties such as the location of the source on the electromagnetic radiation spectrum, useful life, heat generation, level of illumination energy, coherent structure, and compatibility with human operators in the work cell. Significant characteristics of the more frequently used sources are described next:

- *Incandescent bulbs:* This most commonly used light source includes devices ranging from standard household bulbs with reflectors to high-power quartz halogen lamps. Fiber optic bundles are often used to pipe the light onto specific locations on the part. Two major considerations are bulb life and removal of the heat generated by the bulb.

- *Fluorescent tubes:* The reduced infrared energy (heat) produced by this type of illumination, plus the extended life of the tube, make it more efficient than the incandescent bulb. The natural diffused light produced by fluorescent tubes is preferred for vision applications with highly reflective parts.

- *Xenon flash tubes:* When used as a strobe light source, the xenon flash tube is an important light source in vision applications. Xenon strobe light is used when the vision system must capture the image of a moving part. The 5 to 200 microsecond flash of light illuminates the part at one point in its motion. Since CCD vision cameras are temporary storage devices, the image of the part created by the flash is stored by the camera and the part features are scanned into the vision system memory. Two additional advantages of xenon tubes are light with frequencies from a broad part of the spectrum and high-intensity levels.

- *Lasers:* The laser is an important source because it produces coherent light. Coherent light does not disperse as it travels from the source to the target. As a result, the diameter of the laser beam at the target is very close to the beam size leaving the laser. This condition makes the laser a good source for structured light applications.

The selection and design of the light source is a critical part of the vision application. A successful vision application must put an equal emphasis on part lighting and selection of image-capture hardware and software.

6-4 MATERIAL HANDLING

Raw material and parts must be delivered to the automated work cell, and the finished parts must be removed. Material-handling systems are responsible for this transfer activity. The transfer mechanism used to move parts between work cells and workstations has two basic functions: (1) move the part in the most appropriate manner between production machines, and (2) orient and position the part with sufficient accuracy at the machine to maximize productivity and maintain quality standards.

Automated Transfer Systems

The transfer mechanisms used to achieve these two functions are grouped into three categories: *continuous transfer, intermittent transfer,* and *asynchronous transfer.*

Continuous transfer.   In a continuous transfer operation the parts or material move through the production sequence at a constant speed. The work does not stop at a workstation so that a production operation can be performed. If the workstation includes a robot, then the robot must perform the work while the part is moving through the cell. In the manufacturing area a prominent example is the assembly of cars on an automotive assembly line. The cars move at a constant rate through 95 percent of the assembly process, with the automated assembly machine and manual operations performed on a moving vehicle. The picture in Figure 1–4 shows a robot installing front windshield glass on cars as they move through the assembly process. In this case, the robot is capable of moving the glass at a rate synchronized with the material-handling system carrying the car.

Three types of mechanisms are used to achieve continuous motion in manufacturing:

1. *Overhead monorail:* This implementation usually consists of a continuous series of interconnected hooks attached to free-turning rollers that ride in overhead tracks. Parts are hung from the hooks to transport material between workstations or through a production operation. This type of system is used in Figure 9–15 to bring parts past a robot paint sprayer as a painting operation is completed.

2. *Monorail tow system:* This system is similar to the overhead monorail except that wheeled carts are attached to the hooks, and the overhead system pulls the carts from one destination to another.

3. *In-floor tow system:* In an operation similar to the monorail tow, the towing mechanism is a continuous chain riding in a channel placed below the surface of the floor. The chain has hooks placed at regular intervals which pull a wheeled carrier through production. Such a system is the most popular method used to tow car carriers through the main automotive assembly process.

Intermittent Transfer.   The intermittent, or synchronized, transfer system has the following characteristics:

■ Workstations are fixed in place.
■ The motion of the transfer device is intermittent or discontinuous, so that the parts cycle between being in motion and being stationary.
■ The motion of the parts is synchronized so that all parts either move or remain motionless during the same time frame.

Asynchronous Transfer. The term *asynchronous* means *not synchronized;* therefore, an asynchronous transfer system is one in which each part moves independently of other parts. The asynchronous transfer system is often called a *power-and-free system* to indicate that parts can be either free from the transfer mechanism or powered by the transfer device. The robot gripper in Figure 4–18 is poised above an asynchronous conveyor system. The two black strips in the figure form a flexible belt conveyor on which metal pallets ride. The pallets are contained by the metal edges on the outside of the belts. A part of a pallet is visible on the top right of the figure. Pallets are moved by the friction between the belt and the metal pallet base. When the parts on the pallet are needed, the pallet is stopped by a plunger, which rises up between the belts (see Figure 4–18). The belts continue to move, and the pallet slides on the belt surfaces as the robot unloads the parts. Review Figure 3–6 and note that the telephone assembly system also uses a power-and-free type of pallet conveyor system.

The asynchronous transfer system performs more than 60 percent of all material-handling functions in automated production systems.

6-5 PART FEEDING

Recall that the second function of material handling is to orient and position parts with sufficient accuracy to maximize productivity and maintain quality standards. In most automated cells this task is the function of part orienters and feeders. The number of different types of part feeders is as varied as the types of parts used in products; however, several devices are standard in most automated cells.

Gravity Feeders. In the configurations of feeders in Figure 6–8, gravity provides the force to move parts to the pickup point. The configurations in the figure are just a sample of hundreds of styles used to feed single or multiple parts to a manufacturing system for automated handling. The advantage of gravity feeders is the relatively simple design, and the primary disadvantage is that they do not hold a large supply of parts.

Tape Feeders. Tape feeders have components trapped between two layers of tape with equal spacing and a fixed orientation. The tape feeders are clearly visible in Figure 3–6; note the rolls of components at the rear of the feeder. The feeder moves the part to a fixed point and removes the top layer of tape so that the component is ready for automated handling. Tape feeders are used most frequently with small components that are not suited to vibratory feeding.

Waffle-Tray Feeders. A waffle-tray feeder is usually a plastic tray with equally spaced compartments, each containing a single component for retrieval by automatic part-handling equipment, such as robots. An example of a parts tray is visible on the right side of Figure 3–6 under the robot arm.

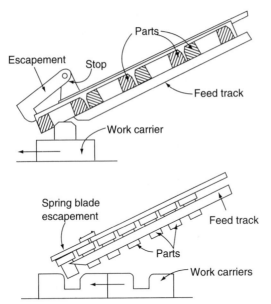

Figure 6-8 Gravity Feeders.

Vibratory Feeders. Two types of vibratory feeders are common: linear track and bowl. In each case the parts holder is isolated from the feeder base by rubber pads, and the holder is vibrated at a variable rate to move the parts. Track vibrations are along the length of the track, and the vibration of the bowl is rotary about the centerline of the bowl. Three bowl feeders are visible in Figure 3–6; the different bowl sizes are dictated by the size of the component part. Bowl feeders have the parts placed randomly at the bottom of the bowl. The vibrating motion causes parts to move up the circular ramps around the inside of the bowl (see Figure 3–6). When the parts are near the top, tabs on the ramp push parts that are not in the correct orientation off the ramp, and the rejects drop to the bottom of the bowl. Bowl feeders are also visible in Figures 1–7 and 2–23.

6-6 INSPECTION

The use of CIM-based work cells demands that quality be verified at each manufacturing step; as a result, the term *quality at the source* was coined. Quality takes many forms, because quality is everyone's job. Quality must be designed into the product and into the system used to make the product. Every person and every automated machine has the responsibility to guarantee that a quality product is produced. To achieve this goal, the production standard is a product free of defects, and the product must be checked frequently to verify that it is free of defects.

A large number of different devices are available to verify that no defects are present in a product. The devices include manual inspection gages and measuring instruments and automated inspection systems. Automated systems range from manual measuring instruments with digital outputs to coordinate measuring machines (CMM). The CMMs are programmed to measure part parameters with little or no human interaction. In many applications, the quality data are transferred to a computer in the work cell through a digital interface present on the measuring device. Because of the computer interface, product quality can be analyzed quickly and saved as a part of the product database.

6-7 AUTOMATIC TRACKING

Tracking the movement of material, parts, and tools through production is a major task. The primary technique used to track the movement of parts and material is *bar coding*, with more than 50 different coding systems in use. Bar code scanning is rapidly replacing the keyboard and other data entry devices for recording manufacturing information and data.

A bar code is a symbol composed of parallel bars and spaces with varying widths. The symbol is used as a graphical code to represent a sequence of alphanumeric characters, including punctuation. For example, the three bar codes pictured in Figure 6–9 represent the numbers written below the codes. The codes can also represent combinations of letters and numbers. The advantages offered by the various coding systems include robustness of character sets; ease of printing and reading; and the ability to represent large numbers with a small code. The two most popular codes in industry are called the *interleave two of five* and *code 39*.

Figure 6–9 Sample Bar Codes. (*Source:* Rehg, James, A., *Computer Integrated Manufacturing.* Englewood Cliffs, NJ: Prentice Hall; 1994, p. 396)

The interleave two of five is a numeric code, so only numbers can be represented by the code. The code was designed with two key features: (1) the code has a high density so that a large amount of information can be encoded in a short space; and (2) the code has a wide tolerance for printing and scanning. This code is a continuous code, meaning both bars and spaces are coded. The odd-numbered digits are represented by the bars and the even-numbered digits are represented by the spaces. The specification requires that the code represent an even number of digits. To satisfy this requirement, a zero is placed in front of all numbers with an odd number of digits before converting to the interleave code. The interleave two of five is effective for coding information on corrugated boxes. In this application, bars must be wide to permit automatic scanning while the boxes are moving on conveyors; however, space on the boxes is usually limited. The code is used widely in the automotive industry and also in warehousing and heavy industrial applications.

Code 39, also called the 3 of 9 code, is the most widely used and accepted industrial bar code symbology. The code can represent the twenty-six letters of the alphabet, the ten digits, and seven additional characters. The code, which can vary in length, has unique start and stop bits. All characters are self-checking, and the intercharacter space does not contain information. Major advantages of this code are the high level of data security provided by the self-checking feature and the wide tolerance for printing and scanning. For example, with good dot-matrix bar code printers and properly selected bar code scanners, the error rate is less than one character substitution in 3 million scanned characters. If high-quality printers and scanners are used, the value drops to one substitution error in 70 million scanned characters.

The scanners used to read bar codes have three major components: (1) a light source (laser, light-emitting diode, or incandescent lamp) to illuminate the code; (2) a photodetector to sample the light reflected from the code; and (3) a microcomputer to convert the photodetector output into the series of letters and numbers represented by the code. Scanners fall into three categories: *handheld contact type, handheld noncontact type,* and *fixed-station systems.* The scanners connect to either portable data storage devices or computers, or they transmit the data over radio frequency links to a host computer.

In many tracking applications the environment does not permit bar codes to be applied or read. For example, the cutting tools used on CNC machines are often covered with cutting fluid; as a result, a bar code attached to the tool would not work. A common tracking technique for items that cannot use bar codes is the *radio frequency tag.* Tags are passive electronic circuits that transmit a code when subjected to radio frequency energy. The passive electronic circuit, shaped like a small computer chip, is held in a slot in the tool base by epoxy. Each circuit transmits a unique code when the tracking device focuses radio frequency energy at the tool. The tracking device receives the tool code and records the number.

Another tracking technique, frequently used on production pallets, uses a binary code to track objects. The pallets have a sequence of holes placed somewhere on the surface. At preselected points along the production process the holes are aligned with proximity sensors or limit switches, with a sensor or switch over each hole. To give a pallet a unique code some combination of holes are filled with pins. When the pallet passes under the sensors or switches, the holes with pins trip the sensor or switch.

Tracking techniques are also used to direct the manufacturing process for a product. In an application at Allen Bradley, for example, a bar code placed on the base of a blank motor contactor tells the production system what type of contactor to build. At every step in the totally automated process, machines add the correct parts to the contactor based on the information read from the bar code. Tracking of material, tools, and products is critical in manual and automated systems.

CASE: CIM AUTOMATION AT WEST-ELECTRIC

Bill looked at his watch. It was almost 7:30—an early start for what could be a long day. The team was meeting at 9 and he had booked the conference room for the entire day. He had just enough time to get the material organized and copies made. Bill walked into the meeting a few minutes after 9; discussion on the design was already underway. After handing everyone a folder, Bill said:

> As you can see we have a long agenda. I would like to start with some of the components of the cell, then move to cell layout. After that we need to work out the number and types of sensors needed. Before we start, I want to mention that Tom O'Brien from cost accounting and I took a rough look at the economic justification. We were conservative with the numbers, but the payback still looks good at this point.

As Bill was about to speak, Marci said, "Ted and I have been looking at the design of the automatic gage. It may be more than our design area can handle. The fixturing gets complex fast because we have to measure the radius on five different-size parts." She paused, then said, "I think we need to go outside to people who build this type of equipment. It will cost more, but I think we will get better results. What will that do to the justification?"

Before Bill could reply, Ted added, "I talked to a couple of vendors to get a ballpark figure. If we are willing to do a little setup on the fixture when we change the dies, then it looks likes $10,000 will get it built." Ted waited for Bill to finish writing, then continued, "That includes a data concentrator which means we have only one serial interface to deal with, and we can download the quality data whenever our controller is ready."

Sensing Ted was finished, Bill said:

> The cost should not be a problem. Ted, would you send them samples of our upset parts and some operational specifications and get a cost figure? Don't make any commitments, but I think that may be the best approach. Before we

tackle the sensor selection problem, let me summarize where we are in the cell design process.

The weight of the fingers, parallel gripper, and heaviest part exceeded the payload capacity of the IRB 2400 robot that we want to use for this application. The robot is rated at 22 pounds, and I want our payload to stay under 15 pounds. Ted is working on a redesign on the fingers. The version in your folders is light enough, but we don't have the finite-element analysis data yet to know if it will hold up. We may have to use composite material for the base of the fingers if this design doesn't have the necessary rigidity.

Several of you have worked on the parts feeder and orienter problem. The parts can be oriented easily with off-the-shelf systems, so we decided to continue to use the parts bins to move the extruded parts to the automated cell. As a result, we've settled on a bowl feeder that will work for all five part sizes to get the orientation needed by the robot. The bowl feeder will be fed by an elevating hopper feeder from a storage bin that holds about 700 parts. The parts will leave the bowl feeder small end first and be held in a vertical orientation on two rails (Figure 5–14j and Figure 6–8). I put a rough sketch of the feeder system in your folders (Figure 6–10). Ted and Marci are working with a vendor and it looks like this part of the cell doesn't present any problems.

After Jerry and I reviewed the oven modifications, we changed the number of slots from 30 to 45. This will permit increased productivity in the future. I included a drawing of the proposed table (Figure 6–11) for everyone to

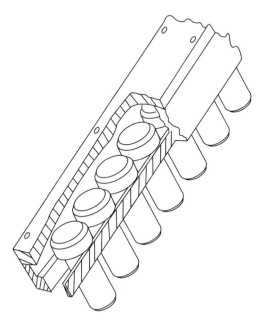

Figure 6–10 Extrusion Feeder System.

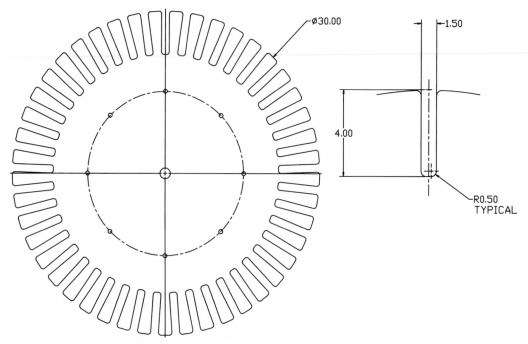

Figure 6–11   Revised Oven Parts Table with 45 Slots.

review. We are still working on a table material that will not warp with repeated heating and cooling. With a robot loading and unloading the slots, we cannot afford to have more than a half-inch variation in the vertical position of the slot as the table rotates through 360 degrees.

During the remainder of the meeting the team worked on the selection and placement of sensors in the cell according to the guidelines Bill developed. A chart (Figure 6–12) was created that included a sensor identification number, the work-cell condition to be sensed, the type of sensor to be used, and comments regarding the operation of the sensor. In addition, sensor numbers were added to the work-cell layout (Figure 6–13). When the sensor selection for the production problems was complete, Marci asked about sensors to detect if someone entered the cell while the robot was operating. Bill responded:

> W-E has a corporate team developing safety guidelines when robot automa-tion is present in a work cell. Let's wait until we get more information on the safety standards we need to follow before tackling the safety sensor issues. Well, the cell is taking shape. Does anyone have any questions or concerns?

After a short pause, Bill said, "Work on the detailed designs in each of your areas and forward the results to me. We'll meet again in a few days. ●

Identification number	Parameter sensed	Sensor type	Comments
1, 2	Oven door position	Limit switch	The limit switch lever arm rotates when the oven door is in the full open position.
3, 4	Oven part table position	Absolute encoder	A digital encoder attached to the oven table shaft provides an absolute binary number indicating the position of the table.
5, 6	Robot shoulder axis position	Proximity	The sensor output changes when the robot arm is fully retracted from the oven.
7, 8, 9	Press upper die position	Photoelectric	The sensor output changes when the upper die of the press is in the full up position.
10, 11, 12	Finished part ejected from the forge	Photoelectric	The sensor output changes when the air-ejected part from the press slides down the finished part chute.
13, 14	Maximum desired oven temperature	Temperature switch	The sensor contacts close when the oven temperature exceeds the maximum allowed level.
15, 16	Minimum desired oven temperature	Temperature switch	The sensor contacts close when the oven temperature falls below the minimum allowed level.
17, 18	Oven door position	Limit switch	The limit switch lever arm rotates when the oven door is in the closed position.

Figure 6–12 Work-Cell Sensors.

Identification number	Parameter sensed	Sensor type	Comments
19, 20	Part oriented and in correct position	Photoelectric	The sensor output changes when a part is in position and ready for the robot.
21, 22	Gripper finger separation distance	Proximity	The output of the proximity switch on the robot gripper indicates that the grippers are closed but no part is present.
23, 24	Parts level in bowl feeder	Proximity	The sensor output changes when the level of parts in the bowl falls below a minimum level.
25, 26	Number of parts in feeder chute	Photoelectric	The sensor output changes when the number of parts remaining in the feeder chute falls below a minimum level.
27, 28, 29	Press cycle	Limit switch	The plunger on the limit switch has not cycled when the press fails to complete a forge cycle.
30, 31, 32	Hot part in lower die	Photoelectric	The sensor output changes when the radiation from the hot part triggers the sensor input.
33, 34, 35	Hot part is fully seated in die	Photoelectric	The sensor output changes when the part is fully seated in the die hole.

Figure 6–12   *(Continued)*

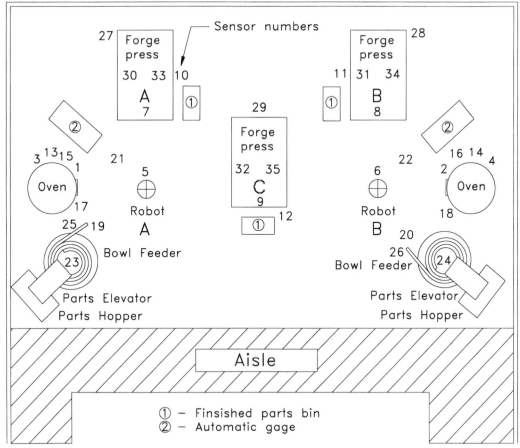

Figure 6–13 Sensor Locations in the Upset Automation Work-Cell.

6-9 SUMMARY

Vision is used to identify parts, determine part position and orientation, inspect parts, and find ranges for three-dimensional location problems. Vision systems are available in three architectures: stand-alone systems, systems based on a programmable logic controller, and integrated systems. In each case, the basic system components include a lighting system, one or more cameras, camera controller, interfaces to external equipment, video image monitors, and programming devices. Most vision systems use the charged-couple device camera; when larger images of surfaces are needed, a vidicon tube is employed.

The basic unit of image measurement is the gray scale, the basic parameter is light intensity, and the basic measurement element is called a pixel. Pixels are arranged in either a linear or rectangular array, and the output of each pixel varies between a fully *on* condition and *off* state as a result of light from the image. The image of the part that is captured by the camera must be analyzed. The first step in analysis is identification of the part edges and features using either edge detection or clustering. Identification of the captured image involves three techniques: template matching, edge and region statistics, or statistical matching using an algorithm.

Lighting for machine vision is a critical element because the cost of the vision system and the success of the application are directly affected by the techniques used to light the part. Three factors affect the selection of the lighting source: (1) the type of features that must be captured by the vision system; (2) the need for the part to be either moving or stationary when the image is made; and (3) the degree of visibility of the environment in which the image must be captured. The light source is applied to the application by means of three techniques: front lighting, backlighting, and structured lighting. Front lighting is used most frequently. If objects are in motion during the image capture time, then a strobe light is appropriate. The illumination sources in vision applications include incandescent bulbs, fluorescent tubes, xenon flash tubes, lasers, infrared light-emitting diodes, and X-ray tubes.

Material-handling systems are responsible for two basic functions: moving parts in the best manner and orienting parts for easy handing. The transfer mechanisms used to achieve these two goals include continuous transfer, intermittent transfer, and asynchronous transfer. In a continuous transfer operation, the parts or material move through the production sequence at a constant speed. The types of mechanisms used to achieve continuous motion in manufacturing include overhead monorail, monorail tow systems, and in-floor tow systems. The second type of material movement system is called intermittent, or synchronized, transfer. This transfer system has fixed workstations, and all the parts on the transfer device stop in unison at the work cell. The third type of material movement system is called asynchronous transfer. The asynchronous transfer is often called a power-and-free system to indicate that parts can be either free from the transfer mechanism or powered by the transfer device.

In the second function of material handling, orientation and positioning, a variety of part orienters and feeders are used. The standard devices include gravity feeders, tape feeders, waffle-tray feeders, and vibratory feeders.

Quality takes many forms, because quality is everybody's job. Every person and every automated machine has the responsibility of guaranteeing that a quality product is produced. The devices that check for a quality product include manual inspection gages and measuring instruments and automated inspection systems called coordinate measuring machines.

Tracking the movement of material, parts, and tools through production is a major task. The primary technique for tracking the movement of parts and material is bar codes. A bar code is a symbol composed of parallel bars and spaces with varying widths. The symbol is a graphical code that represents a sequence of alphanumeric characters plus punctuation. The two most popular codes in industry are called the interleave two of five and code 39. The scanners that read bar codes can be grouped in three categories: handheld contact type, handheld noncontact type, and fixed-station systems. A common tracking technique for items that cannot use bar codes is radio frequency tags. The tags are passive electronic circuits that transmit a code when subjected to radio frequency energy. Another tracking technique, common on production pallets, utilizes a binary code pattern generated by mechanical switches or electronic sensors.

QUESTIONS

1. Describe five application areas for vision systems.
2. What are the basic components of a vision system?
3. What two types of cameras are currently used?
4. Describe the operation of a solid-state CCD camera and define the terms *pixel* and *gray scale*.
5. Describe edge detection and clustering.
6. Describe the three commonly used techniques for two-dimensional recognition with vision systems.
7. Describe the three factors used to select lighting for vision applications.
8. Describe the three lighting techniques used in vision applications.
9. What conditions in the application require the use of strobe lighting?
10. What are the advantages and disadvantages of the six lighting sources?
11. What are the two basic functions of material handling?
12. Describe the three basic types of material-handling systems.
13. Describe the four standard part feeders used most often in automated cells.
14. What is a CMM and how is it used?
15. Describe the three techniques most frequently used for automatic tracking of production parts and material.
16. Compare and contrast the interleaved two of five and code 39 bar codes.
17. List and describe the major components of bar code scanners.

CASE PROJECTS

1. If the gripper base in the W-E case weights 5 pounds, what is the total weight of gripper, fingers (original design in stainless steel), and largest part?
2. If stainless steel fingers are used and the total payload weight is 15 pounds, what is the maximum volume for each finger of the gripper?
3. Categorize the sensors selected by the W-E team in the case into the six basic reasons why sensors are used. Are there any other sensors they should have included?
4. Categorize the sensors selected in question 4 of the case/projects section in Chapter 5 into the six basic reasons why sensors are used. Are any areas not covered?
5. If the part hopper holds 700 parts and the work-cell cycle time is 12 seconds per part, how long can the cell run unattended?

6. With a 45-slot carousel and a required oven time of 450 seconds, what is the maximum number of parts that can be manufactured in a week from the automated cell?

7. Analyze the quality requirements for the blade production line and identify areas where vision could be used for inspection.

8. In addition to the vibratory feeder, what other techniques for part orienting and feeding could be used in the blade production line?

9. How could bar code technology be used to track the blade production process?

Chapter

7

Robot and System Integration

7-1 INTRODUCTION

The CIM wheel introduced in Chapter 1 (Figure 1–5) identified the major areas present in manufacturing enterprises and indicated the interdependence and operational integration that must be present. The integration of information among the elements in the CIM wheel is illustrated by the enterprise network architecture in Figure 7–1. Note that information flow in the three major elements—product and process definition, manufacturing planning and control, and factory automation—is supported by local area networks (LANs). Network communication between the major areas occurs over the enterprise network backbone.

The CIM architecture for an enterprise is built around a common central database, or single image of the product, and production data that support the needs of every user. The generation of product data begins with the design of the product. The part specifications and drawings and lists of material for the product are often the first additions to the product database. The design data are used by many other departments; for example, purchasing uses the list of material, called the bill of materials, to purchase parts that are not manufactured in house. The production planning area uses the part drawings to plan the manufacturing sequence. In the past, the design information was given to other departments through paper copies and interoffice mail. It was difficult to keep all of the distributed documentation current because design updates occur throughout the development of a product. As a result, sometimes parts were ordered that were no longer necessary, or planning was done on parts that had changed significantly. The CIM architecture avoids these problems by keeping a single image of the product data available to

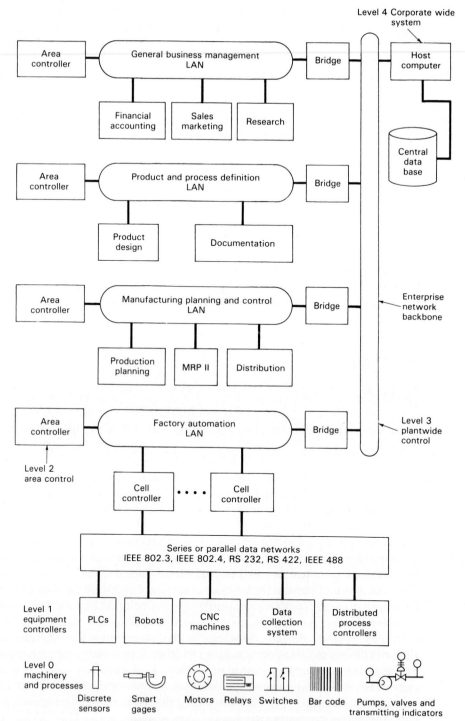

Figure 7–1 CIM Computer Architecture.

all departments. For example, the bill of materials is entered into the manufacturing planning and control software system from the design and updated as changes occur. Therefore, the current parts list, stored in the central database (Figure 7–1), is available to anyone on the enterprise network.

The architecture also provides a core of common services for the central data: communication and distribution of data, management of the data, and presentation of the data. The data take three distinct forms: (1) alphanumeric characters called *text* data; (2) graphics data called *vector* data; and (3) bit-mapped scanned pictures called *image* data. Each data type presents a unique integration problem for the three core services just described. The degree to which the CIM architecture performs the three core services on all data types is a measure of the quality of the CIM implementation. Therefore, the general CIM architecture includes the physical system and the enterprise data. The physical system is divided into hardware and software. The hardware includes the computers, input devices, output devices, physical networks, and network interfaces. The software, as the name implies, is the software used throughout the system.

The following sections describe the hardware and software used to integrate the shop floor with the enterprise database and to control the operation of computer-driven automation in production cells and systems.

7-2 SYSTEM OVERVIEW

The primary focus of this text is the integration of robots into automation cells and systems. Nevertheless, an understanding of the larger enterprise-wide information system is essential for an understanding of the detailed operation of the automation cells. The next two sections provide a general overview of the hardware and software that support the enterprise network illustrated in Figure 7–1.

Hardware Overview

The hardware is distributed across the enterprise, and as Figure 7–1 illustrates it includes host, area, cell, and production machine computers; computer peripheral devices; different types of networks; and network interface devices. Users often have microcomputer workstations networked to the system at the area level. The most frequently used network hardware is *ethernet* and *token ring*, with the former found in more shop floor applications. If both networks are used in the CIM system, then a gateway computer is required to interface the two networks.

Host computers are usually mainframe machines with large data storage capability. The area-level computers are frequently midsized machines, like the IBM AS400 or the DEC Alpha family, that run local user applications and act as data concentrators. In some smaller installations microcomputers are used as area controllers. The cell controllers are either industrial microcomputers like the industrial PC or Gear Box from IBM or a smaller DEC MicroVAX or Alpha system. The computers used for production machine control vary widely. Included in the group are microcomputers, microprocessors, and programmable logic controllers.

Software Overview

The software distributed across the enterprise in a CIM implementation takes many forms. Each computer from production machine controller up to the host has an *operating system*. Examples of operating systems at workstation or cell-control level include DOS, OS/2, UNIX, and Windows NT. OS/2, UNIX, and Windows NT support multiple sessions so that the user can have two or more programs executing and present on the monitor simultaneously. All the computers in the enterprise that share the network must have *network software* running. The network software permits communication between active computer nodes in the enterprise system. For example, a cell-control application program could request a different robot program file from the host.

All the computers in the CIM network have one or more application programs running under the operating system. Using network software the applications can access data from other computers in the system. The applications are as varied as the departments in which they are used. The front office can be using word processing, desk-top publishing, and electronic mail, whereas the product design department can be using CAD, finite-element analysis, word processing, and electronic mail. The marketing department can retrieve product drawings from the central database that were developed on the CAD computers in the design department.

7-3 WORK-CELL ARCHITECTURE

The architecture of automated cells includes layers of computer control. The area controller is at the top and smart, or microprocessor-controlled, devices are at the bottom. Often the architecture includes programmable logic controllers for sequential control of the cell and some production machines, robot controllers to drive the servo robots, and computer numerical control (CNC) for process machines. Commonly used cell control devices are described in the next sections; then an example of the architecture used in a machining and assembly cell is presented.

Cell Controllers

The cell controllers are either industrial computers that use standard microprocessor chips or minicomputers with proprietary processors. For example, the IBM Gear Box (Figure 7–2), an industrial computer designed for operation in harsh production environments, uses an 80486 or Pentium microprocessor. Minicomputers from Digital Equipment Corporation (DEC)—some of the most frequently used machines for area and cell control applications—use a proprietary chip called Alpha (produced by DEC). The system configuration for cell controllers includes internal hard drive and floppy drive for data and program storage, internal memory (RAM), keyboards, pointing devices, and monitors.

Cell Control Software Structure

The operating system software used for cell controllers in most applications is OS/2, Windows NT, or UNIX. These operating systems can concurrently run

Figure 7–2 IBM Industrial Computer. (*Source:* Rehg, James, A., *Computer Integrated Manufacturing.* Englewood Cliffs, NJ: Prentice Hall; 1994, p. 376)

multiple application programs. This multitasking feature permits the cell controller to have multiple application programs running at the same time. For example, the cell controller can execute a program to collect quality data from a smart gage while another program downloads a program to a CNC production machine to cut the next part. Work-cell controller applications that demand this level of flexibility require multitasking operating systems.

The cell controller interfaces with other intelligent machines and devices at different levels in the control architecture using standard serial data interfaces and networks. For example, a cell controller is often attached to the factory LAN (Figure 7–1) and interfaced with devices in the cell. Figure 7–1 indicates that the interface to work-cell devices uses a variety of serial and parallel data networks: IEEE 802.3 (ethernet), IEEE 802.4 (token ring), RS 232, RS 422, and IEEE 488 general-purposes instrumentation bus (GPIB). A typical software configuration in the cell controller to manage these many interfaces is illustrated in Figure 7–3. Study the figure until you are familiar with the terms in each box.

The operating system (OS) is usually one of the three listed at the bottom of the figure. The application program interface (API) and system enabler (SE) software resides between the OS and the applications. An API, such as IBM's Distributed Automation Edition, handles some of the operational overhead that is common to all applications; in addition, an API helps manage the resources, such

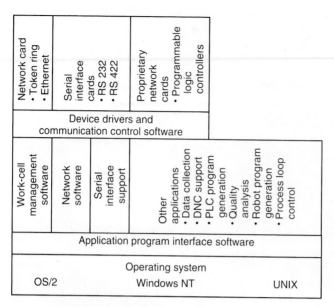

Figure 7–3 Work-Cell Controller Software Layers. (*Source:* Rehg, James A., *Computer Integrated Manufacturing.* Englewood Cliffs, NJ: Prentice Hall; 1994, p. 378)

as the hard drive data storage, across all the computers on a network. The use of API software permits the developers to spend more time programming the specific application and less time programming routines required by the operating system. The applications can include a wide variety of software (see Figure 7–3).

The cell controller frequently has different interface cards to support network standards like the ethernet and token ring. The device drivers and communications control software reside between the interface cards and the application software (see Figure 7–3). In some applications, a proprietary interface card is installed to permit direct communication between a work-cell device, like a programmable logic controller or robot, and the cell controller.

Work-Cell Management Software

The cell management software used in cell controllers manages all of the cell activity. The primary role of the cell controller is communication and information processing. For applications higher in the architecture, the cell controller is a concentrator of information generated in the cell and a link between production machines and applications running on computers in the design and the production control areas. On the downstream side, the cell controller is a distributor of data, information, and program files to the computer-controlled devices of the work cell. The cell controller interfaces with production machines and transfers data files as small as a single bit to files containing megabytes of data. Other cell control responsibilities include management of program libraries for devices in the cell, support for engineering change control, and production tracking. The information and communication tasks frequently consume 80 to 90 percent of the resources of the processor in the cell controller.

Applications typically loaded into the cell controller include these:

- Production monitoring
- Process monitoring
- Equipment monitoring
- Program distribution
- Alert and alarm management
- Statistical quality and statistical process control
- Data and event logging
- Work dispatching and scheduling
- Tool tracking and control
- Inventory tracking and management
- Report generation on cell activity
- Problem determination
- Operator support
- Off-line programming and system checkout

7-4 PROGRAMMABLE LOGIC CONTROLLERS

The *programmable logic controller* (PLC) was developed in the early 1970s for General Motors. The early PLCs were special-purpose industrial computers designed to eliminate relay logic from sequential control applications. Current PLCs are able to control discrete and analog processes, attach to general-purpose and proprietary LANs, provide servo and sequential control for robots, and serve as the primary controller for cells or complete automated systems. An understanding of robot automation is not complete without an understanding of PLC operation and programming. The following sections give an overview of PLC operation, and programming is described in Chapter 8.

PLC System Components

The PLC is a computer designed for control of manufacturing processes, assembly systems, and general automation. The mechanical configuration of PLCs includes a *rack* into which *modules* are inserted. The picture in Figure 7–4 shows the rack of an Allen Bradley PLC with modules, a processor, and power supply installed. The modules and processor are inserted into electrical connectors on the back plane at the back of the rack. The back plane contains the power and signal conductors to interconnect all the modules that are inserted into the rack. The modules generally fall into one of the following categories: interfaces to production equipment, network and serial communications, special-purpose modules, power supplies, and a microcomputer. A description of systems using each of these modules is included in the next section.

Figure 7–4 Allen Bradley PLC Rack and Modules. (*Source:* Rehg, James A., *Computer Integrated Manufacturing.* Englewood Cliffs, NJ: Prentice Hall; 1994, p. 385)

Basic PLC System Operation

The operation of PLC components, for the control of automated systems, is described in the block diagram in Figure 7–5. Study the figure and note the components that are part of the PLC (inside the dashed box) and those that are external to the PLC.

The computer, called the *processor* in PLC systems, is at the heart of the PLC operation. *Input modules* receive electrical signals (Figure 7–5) from a wide variety of sensing devices and outputs from other systems. The source labeled *other outputs* requires some clarification. In complex automated cells with different PLC vendors, the output from one PLC is often connected to the input of another PLC. Under these conditions, the information processed by the first PLC is used by the second PLC to control another part of the process. In that case, the output from one PLC is the input to a second PLC.

The processor performs arithmetic and logical operations on input data and turns *on* or *off* outputs in the *output modules* based on the program resident in the process. The output modules are wired to system components that control the process. The input module, processor, and output module blocks are the only ones

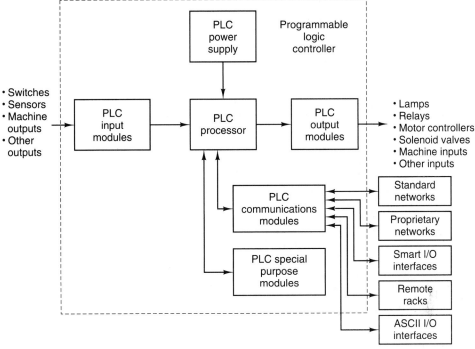

Figure 7–5 Programmable Logic Controller System Block Diagram. (*Source:* Rehg, James A., *Computer Integrated Manufacturing.* Englewood Cliffs, NJ: Prentice Hall; 1994, p. 387)

required for most PLC automation applications. Inputs are scanned and outputs are changed based on the input conditions present and the logic programmed into the PLC processor.

The PLC *communications modules* are not used as frequently as input and output modules; however, communication is a critical part of every robot automation project because production data must be available to departments across the enterprise. The *standard networks* box in Figure 7–5 indicates that the PLC can be placed directly on the factory LAN to communicate with other cell or area controllers. The standard network supported by most vendors is ethernet, and common protocol standards are TCP/IP and manufacturing message specification (MMS). In addition, proprietary LAN software from most major vendors is also available.

The *proprietary networks* box in Figure 7–5 indicates that most PLC systems have a proprietary network available to link PLC processors together in a LAN. In most cases only PLCs from the same vendor are compatible with the network interface and protocol.

The box labeled *smart I/O interfaces* in the figure includes PLC hardware that interfaces smart devices to the PLC through a serial data link. The term *smart* implies that the external device has a microprocessor and can be programmed. For example, operator panels have switches for control of process machines and devices, and the panels have lights to indicate the condition of process equipment. Frequently the operator panels are located in a control room away from the process itself. The traditional approach in building an operator panel requires a minimum of one wire per switch and lamp, plus a number of return wires between the operator panel and PLC input and output modules. As a result, the wire bundle between the panel and the PLC can have hundreds of wires running hundreds of feet in conduit. In contrast, the smart operator interface uses the same number of switches and lamps but controls them with a microprocessor located in the operator panel. The communications interface between the operator panel and the PLC is just a single coaxial cable for the transfer of the data between the PLC and panel processor. A large number of smart external devices are available: motor controllers, process controllers, text readout devices, programmable CRT displays supporting full color and graphics, voice input and output devices, and discrete input and output devices.

In an effort to distribute the control capability across a large automation system, PLC vendors provide *remote rack* capability (Figure 7–5). The rack uses the standard I/O modules for the control of machines and processes; however, the processor module is replaced with a remote rack communications module. The remote rack is controlled by the program and processor in the main PLC rack, and communications between the two racks is supported by a single coaxial cable.

Another communications box illustrated in Figure 7–5 is the ASCII interface. This communication resource is either built into the processor module or comes as a separate module. In both cases the ASCII interface permits serial data communication using a number of standard interfaces (eg, RS232 and RS422).

7-5 COMPUTER NUMERICAL CONTROL

The definition of computer numerical control (CNC) was restricted initially to just numerical control (NC) machines associated with metal cutting. Currently, CNC represents all production machines that use internal computers to control the movement of tooling with production programs. The basic robot system described in Figure 1–9 is an example of a machine that fits the definition of CNC. Consult the generic block diagram for CNC machines in Figure 7–6. Compare the robot controller in Figure 1–9 with the CNC block diagram and verify that the functions are similar.

The input to the CNC system comes from either an operator or from another machine. Frequently the input for CNC machines is another computer. For example, many of the CNC machines received the program they used to cut parts through a serial connection to a computer, such as a cell controller. In other cases, the CNC machines and the cell control computer are connected to a direct numerical control network to download programs to the CNC machines. The production machines using CNC include robots, coordinate measuring machines and most material processing machines.

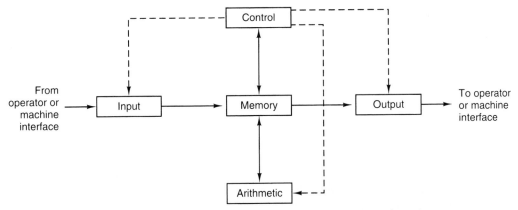

Figure 7–6 Five Major Functional Units of Computer Numerical Control. (*Source:* Rehg, James A., *Computer Integrated Manufacturing.* Englewood Cliffs, NJ: Prentice Hall; 1994; p. 394.)

7-6 CONTROLLER ARCHITECTURE

The controller architecture for servo-type robots is different from that used in nonservo systems. A description of each is provided in the following sections.

Nonservo Robot Controllers

The sequential movements of nonservo robots are controlled by devices external to the robot's arms. The types of controllers used by pick-and-place robots include the following:

- *Drum programmers,* such as the Seiko sequencer illustrated in Figure 7–7, are used in some applications. In this unit, as in most mechanical drum programmers, the operational sequence is set by the arrangement of tabs or cams on the drum surface. The tabs actuate either electrical switches or hydraulic-pneumatic valves that control the movement of each robot axis. The time for each sequence is determined by the number of tabs used and the speed of rotation of the drum.
- *Pneumatic logic* or *air logic programmers* are also used to control the sequences of pick-and-place pneumatic robots. An air-powered logic network is built using fluidic logic elements to provide the sequential control the robot requires. The air logic system is built and programmed by connecting the fluidic elements together by small plastic air lines. The time and sequence of robot movements is determined by the fluidic elements used and the way in which they are interconnected.
- *Programmable logic controllers* are the most frequently used devices to control the sequential movement of pick-and-place machines. Figure 7–8 illustrates the use of a PLC in a typical low-technology robot system. The figure

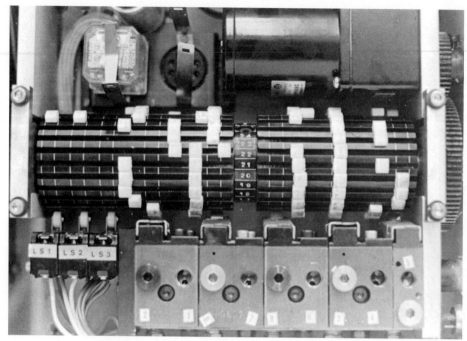

Figure 7–7 Drum Programmer for Seiko 700 Robot.

indicates that the PLC is used for many functions in the cell and as an interface to the CIM system. The interface between a pneumatic robot with three axes and a PLC is illustrated in Figure 7–9. As you trace the pneumatic and electrical circuits in the interface, note that four pneumatic valves control the three axes and gripper, and a fifth valve controls an intermediate stop actuator.

The operation of controllers for nonservo robot systems varies with system size and control technique used. For example, if a drum controller is used the operator would just turn power on after the tabs are placed in the correct position on the drum. In comparison, a system that uses a PLC to operate the robot and other work-cell hardware may require a greater amount of operator involvement. The program may have to be loaded into the PLC from magnetic tape or from a computer, and the operator may have to interact with the production process performed in the work cell.

Servo Robot Controllers

The block diagram in Figure 7–10 illustrates the typical modules found in a controller for a servo-controlled arm (recall the robots described in Chapter 2). The heart of the systems is the CPU, which is responsible for memory management, input/output management, information processing, computation, and control of

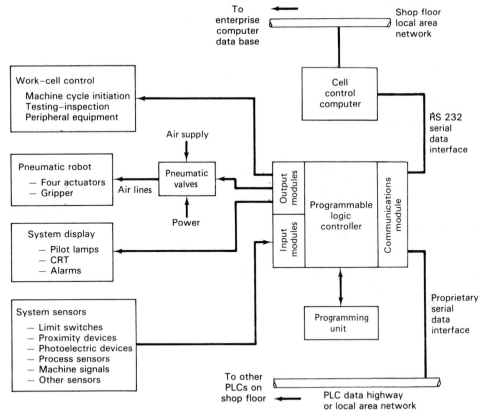

Figure 7–8   Robot System Driven by Programmable Logic Controller.

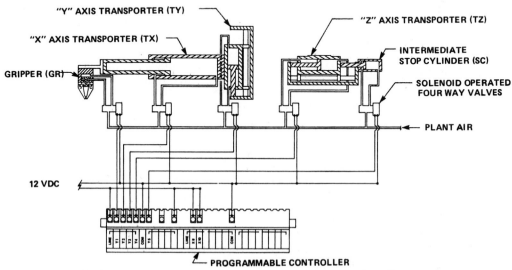

Figure 7–9   Programmable Logic Controller and Robot Interface. (Courtesy of Mack Corporation)

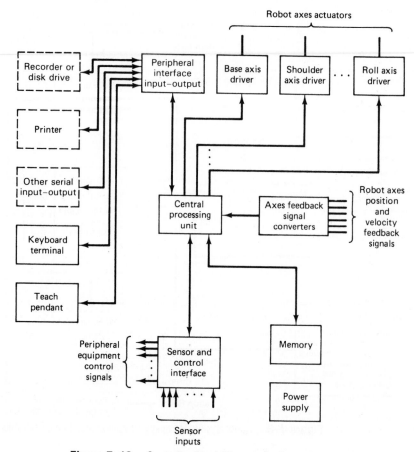

Figure 7–10 Controller Block Diagram for Servo System.

each robot axis. The CPU configurations for current robots demonstrate the alternatives provided by the explosive growth of computing hardware. The options adopted by some robot manufacturers include the following:

- Using off-the-shelf minicomputers, such as the DEC MicroVAX or Alpha series, to configure a special-purpose computer capable of handling robot CPU duties
- Using either a 32- or a 64-bit microprocessor as the base of the CPU design
- Using a network of 16-, 32-, and 64-bit microprocessors linked together by hardware and software to perform the functions of the CPU. Most of the current state-of-the-art machines follow this practice.

The last technique described, networking, offers several advantages for robots in the high-technology group. The system operation is much faster when the duties of the CPU are divided up among different processors.

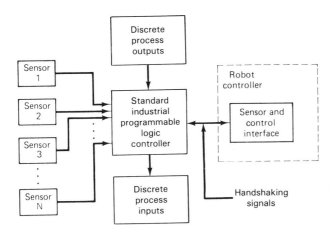

Figure 7–11 Block Diagram of Robot Controller and External Programmable Controller.

Memory. The memory used in most robot controllers is solid-state with battery backup for program protection during loss of primary ac power. The operating system and application program are usually stored in volatile devices (RAM), and nonvolatile devices (ROM) are used to store basic load routines to wake the system up and load the controller programming language.

Peripheral Interface. The CPU controls the interface between two different types of peripheral hardware. The first is digital data communications between the robot controller and the peripheral devices needed to operate the robot system. The most commonly used peripherals are the teach pendant and the keyboard terminal. More complex systems often include a recording device for program storage and playback, a printer for hard copy, and a general-purpose serial port for communications with cell controllers or vision systems. The second type of interface is for discrete or *on/off*-type signals. This interface in the servo-type robot controller monitors the output of sensors in the work cell to determine if a change of state has occurred. For example, a sensor to detect the presence of a part in the gripper would be monitored through an input of this interface. If the gripper acquired the part, the sensor output would change from *off* to *on,* and the controller would drive the arm to the next programmed point only when this input change was received. The output portion of the interface has the capability to control discrete devices. For example, in a punch press loading application the robot program can activate the press to start processing a part that has just been loaded.

The discrete interface section of the robot controller is basically a built-in PLC for use in the work cell. Typical input values of 80 to 140 volts ac and 6 to 40 volts dc are available on the discrete input interface of some robot controllers. The output interfaces will switch either ac or dc currents in ranges that are compatible with the input section. Figure 7–11 shows a standard PLC monitoring work-cell sensors and passing the logical results of their status on to the robot controller.

7-7 INTERFACES

The integration of production cells into CIM systems dictates that robots and automated systems must work with other equipment in the enterprise. If this is so, then robots must be interfaced to the other hardware.

An interface is defined as a place at which independent systems meet and act on or communicate with each other.

In an automated system, there are devices that must be interfaced so that communication between systems is possible. The numerous independent systems are grouped into three interface categories called *simple sensor interface, complex sensor interface*, and *enterprise data interface*. All interface requirements of the automated robot cell fall into one of these categories.

Simple Sensor Interface

Currently the most well-defined interface area is the simple sensor interface. A simple sensor is basically one that has its signal originate in some peripheral hardware or device. Peripheral hardware is defined as equipment used in conjunction with an industrial robot in the design of a work cell. An additional requirement stipulates that the communication with the robot must be by discrete signals only (*on* or *off*). The group does not include peripheral devices such as disk drives, printers, or equipment using higher-frequency binary-coded signals with a data format. Equipment in this group does include all the discrete sensors, such as limit switches, proximity sensors, and photoelectric sensors. In addition, all discrete process signals necessary to control the cell are included. These would be signals from machine tools, welders, and material handlers, plus the signals that the PLC, cell controller, or robot originates to operate these peripherals.

The standard logic signal levels for this interface are 0 volts for the low level, and 110 volts ac or 24 volts dc for the high level. Some robot controllers use logic levels of 0 volts and 5 volts, however, and are identified as *TTL* signal devices. The input/output interface uses optical coupling between the PLC, cell controller, or robot and the peripheral to assure isolation of power and grounds between the different systems. Figure 7–12 illustrates a typical input/output module for a PLC or robot with the typical wiring required. Figure 7–13 shows the opto-isolators provided by the robot controller for input-output interfacing. Note in Figure 7–12 that the only connection between the robot and the external circuit is the light that activates the solid-state switch. The output module acts as a switch to apply power to the external load. The input module has one lead connected to the electrical common and the other lead connected to the external sensor that switches power to the input and activates the light coupling and the solid-state switch. The opto-isolated driver is usually an LED, and the solid-state switch is a light-activated transistor for dc or a triac for ac modules.

Electrical noise from industrial machines and contact bounce on the sensor switch contacts can cause input/output interface problems in the system. Adopting the following six guidelines will eliminate many of the potential problems:

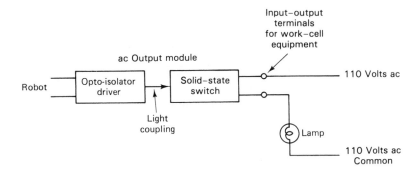

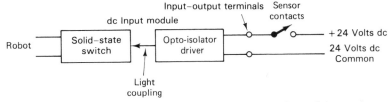

Figure 7–12 Simple Sensor Interface in Robot Controller.

1. Eliminate ground loops and provide good grounds.
2. Shield wires when excessive electrical noise is present.
3. Use damping in the sensor contacts or delays in the PLC or robot program to avoid the multiple signals produced by bouncing contacts.
4. Use only opto-isolated input/output modules.
5. Use arc suppression circuits to reduce noise from switching inductive devices.
6. Route 110-volt ac and 24-volt dc signal lines in separate wire bundles and cable trays.

No standard connector type or size is recommended for the simple sensor interface. Most manufacturers use screw-type terminal strips for termination of signal wires.

Complex Sensor Interface

The complex sensor interface is used with sensors that require some signal conditioning before the data are transferred to the PLC, cell controller, or robot controller. A complex sensor is defined as one that requires some type of preprocessor to perform analog-to-digital conversion, scaling, filtering, formatting, analysis, or coordinate transformation on the raw data before they are presented to a higher level controller. A complex sensor communicates with the complex sensor interface through either digital or analog signals, but the interface communicates with the higher level controller using only digital signals.

Figure 7–13   Input/Output Modules in Robot Controller.

The complex sensor system gathering the data frequently must perform computational operations on the data to do the following:

- Convert from analog values to digital values that are compatible with the higher level controller.
- Detect features or recognize patterns present in the information.
- Compare measured data with values previously stored in the controller or sensor interface.
- Perform coordinate transformations.

Data from vision and torque/force sensors illustrate the way in which complex computation must be performed on the sensory information before it can be used by the higher level controller.

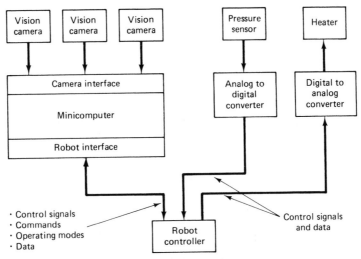

Figure 7–14 Block Diagram of Complex Sensor Interface.

In addition to the computation requirements placed on the complex sensor interface, it must also operate at high speed. The very nature of the job to be accomplished—error correction and trajectory control—requires that the closed-loop response time of the interface be very short. Decisions cannot be made and errors corrected until the data are presented to the higher level controller. Delays of more than a few milliseconds do not permit high-performance closed-loop operation.

Bidirectional data communication is another important characteristic of the complex sensor interface. Not only must the interface talk to the PLC, cell controller, or robot controller, but the PLC and controllers must communicate with the interface. In the most complex interfaces, the two-way communication includes control signals, serial or parallel data, and addressing information. For example, in vision applications the robot controller could signal the vision system when a part needs to be analyzed, and the vision system sends the coordinates and orientation of the part to the controller.

The choice of complex sensor interface is dictated by the type of complex sensory information present. The system may require one sensor per interface if processing speed is important, or it may permit one interface to handle many sensors. At present, vision, torque/force, and remote positioning tables require the most complex interfaces. Figure 7–14 is a block diagram of a robot controller and complex sensor system. In one of the interface modules, the vision sensor requires a minicomputer with memory for storage of data and vision system programs. In another, a pressure sensor needs an analog-to-digital converter for data transformation. In a third, a digital-to-analog converter is used to drive the heating unit. As automated work cells become more complex, the demands on sensor interfacing will increase. In some current work cells, for example, more than one thousand sensors are required for control of the production line system.

Enterprise Data Interface

The enterprise data interface represents the link between the robot controller and other computers in the work cell and CIM architecture. The interface passes blocks of digital information, usually in the form of ASCII files. The interfaces available in robot controllers include *node-to-node* or *direct* communication devices and also LAN capability. The direct communication offered as a standard feature on most robot controllers uses either the RS232 or RS422 standard for serial connectivity. In addition, some controllers offer LAN options for IEEE 802.3 (ethernet standard), IEEE 802.5 (token ring standard), or the MAP/TOP (Manufacturing Automation Protocol/Technical Office Protocol) standard.

The RIA is working on several standards that affect the communications-information interface area. One is the American National Standard, ANSI/RIA R15.04, for Industrial Robots and Robot Systems—*Communications-Information*. It is a companion for the Manufacturing Message Specification standard ISO 9506. A second standard supported by the AIA, ANSI/AIA A15.08, has six components addressing sensor interface issues in the vision area.

7-8 AN INTEGRATED SYSTEM

The work cell in Figure 7–15 illustrates the hardware and software interfaces present in state-of-the-art production systems. Study the system carefully before reading the following description.

Machining Cell

The *machining cell* performs a milling operation on the material delivered to the cell by a material-handling system. The bar code scanner reads the bar code on the stock and passes the value to the work-cell controller that verifies the stock number with the current shop order. Sensors connected to Allen Bradley PLC verify that the mill is ready to receive material, and a signal from PLC starts the pneumatic robot. The Texas Instrument PLC, a controller for the pneumatic robot, executes a program to load the mill. With the mill loaded, the Allen Bradley PLC signals the work-cell controller to download the CNC code to the mill and initiates the cutting cycle. The vision system and smart gage measure the quality of the finished part, and the Allen Bradley PLC reports the results to the work-cell controller. Quality data are collected in a data-base in the work-cell controller and sent to the production control software package on the host at regular intervals. An operator interface with push-button switches and pilot light indicators, interfaced to the Allen Bradley PLC, provides a visual indication of cell activity and problems and permits operator input.

Assembly Cell

The *assembly cell* assembles the machined part with several purchased parts into a complete subassembly. The bar code scanner reads the code on all the parts delivered to the cell. The cell controller verifies that all parts presented to the cell

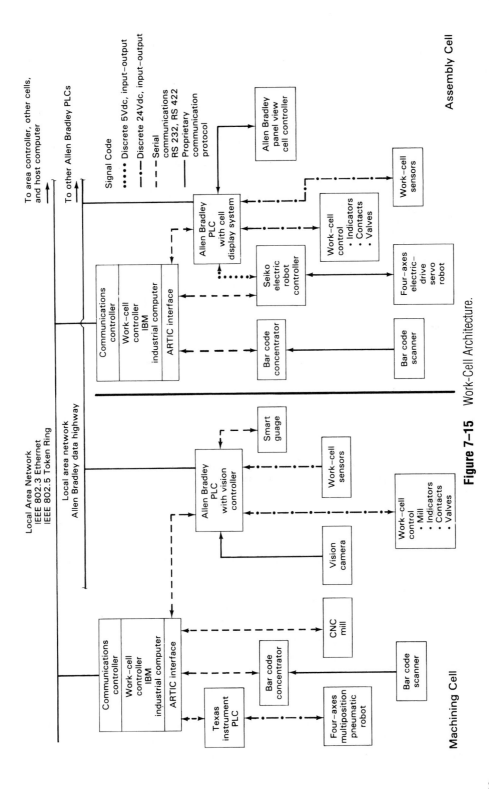

Figure 7–15 Work-Cell Architecture.

are mating parts for the shop order received from the production control software system. The cell-control software sends a value to the PLC that indicates the assembly procedure should start. After verifying from sensor data that the assembly fixture is ready, the PLC sends a 5-volt dc signal to the Seiko robot controller that starts the robot program and the assembly process. During the assembly process the robot controller and the PLC have a preestablished data exchange, called *hand shaking*, that forces the robot to wait for PLC verification of correct sensor data before proceeding to the next assembly step. The *panel view* system attached to the PLC provides a programmable operator interface and dynamic graphic display of cell status and activity.

Both the machining and assembly cells are nodes on two LANs: the enterprise network and the PLC network. Many enterprise networks use the IEEE 802.3 ethernet standard on the shop floor because it has been available for some time.

Signal Types

The connectivity between cell hardware is coded in Figure 7–15 to indicate the type of signal present. The two *discrete* types of signals are either *on* or *off*, with a 24-volt dc change for one and a 5-volt dc change for the other. *Serial data communications* is used frequently in the work cells; RS422 is present most often. Serial data exchange permits large blocks of data, such as programs, to be transferred between computer-controlled machines. Data exchanged between hardware from the same vendor frequently use a *proprietary communication protocol* that is not an industry standard. This type of communication can use either serial or parallel data exchange.

Work-Cell Controller

The work-cell controllers using Pentium or higher microprocessors are microcomputers hardened for the shop floor environment. The work-cell controller could also be a DEC MicroVAX or Alpha for support of cell data communications and control. The IBM PCs have ARTIC (a real-time interface coprocessor) and LAN peripheral cards installed to permit data transfers to the enterprise computer network and the work-cell hardware. The ARTIC is a multiport serial data interface that permits simultaneous control of interfaced hardware. The LAN card using LAN software makes the work-cell controller a node on the enterprise-wide data and communications network.

The software running on the controller includes an operating system, network software, a work-cell control language, applications for the hardware in the cell, and all the necessary ARTIC drivers. The work-cell control language is a software program that permits the cell controller to (1) exchange data (single values or blocks of code) with the devices interfaced through the ARTIC; (2) exchange data (single values or blocks) with computer nodes on the enterprise network; (3) create dynamic graphic screens for the monitor to display work-cell activity and operating conditions; (4) execute mathematical functions using data from the cell and/or evaluate cell data using logical expressions; and (5) initiate cell activity.

Programmable Logic Controller

The PLCs serve several functions: (1) the operations sequencer for all cell activity; (2) the interface to all sensors monitoring cell conditions; (3) the control device for a vision system, smart gage, and work-cell display; (4) the controller for a stop-to-stop pneumatic robot; and (5) a node on a LAN that links all the PLCs on the shop floor. As the description of the cell indicates, the PLC provides for the work cell what a conductor provides for an orchestra; that is, a method to keep every event in the correct order. The Allen Bradley PLC in the machining cell could act as the pneumatic robot controller in addition to controlling the cell activity. It is not uncommon, however, to use a separate PLC, often supplied by the robot vendor, to keep the two programming functions separate.

The system illustrated in Figure 7–15 is an example of work-cell architecture but does not imply a singular solution. The same function could be achieved by many variations in hardware connectivity. In addition, the hardware and software from different vendors could be used to achieve similar results.

CASE: CIM AUTOMATION AT WEST-ELECTRIC

The upset cell design is taking shape. Bill made copies of the upset forging cell control architecture (Figure 7–16 on p. 232) and the preliminary PLC interface drawing (Figure 7–17 on p. 233 and Figure 7–18 on p. 234) and forwarded copies to each team member for comment. ●

7-10 SUMMARY

The robot is a part of the total CIM architecture and must interface with the work-cell hardware and with the enterprise-wide network. The architecture of the enterprise system includes layers of computers ranging from mainframe machines to microprocessor-based production machines. The system is controlled by software operating systems that run application software. The computers used to link the many shop floor functions include area controllers, cell controllers, programmable logic controllers, computer numerical control machines, and robot controllers.

The architecture for the servo-controlled robot systems includes a CPU, memory, peripheral input/output, discrete input/output, power converter or supply, axes drivers, and feedback signal conditioning. Although the nonservo robot system also has these elements, the axes drivers are part of the discrete input/output, and the feedback network is absent. The nonservo robot controller is often manufactured by the vendor who builds the robot arm, but in some cases a standard industrial PLC is used for arm control.

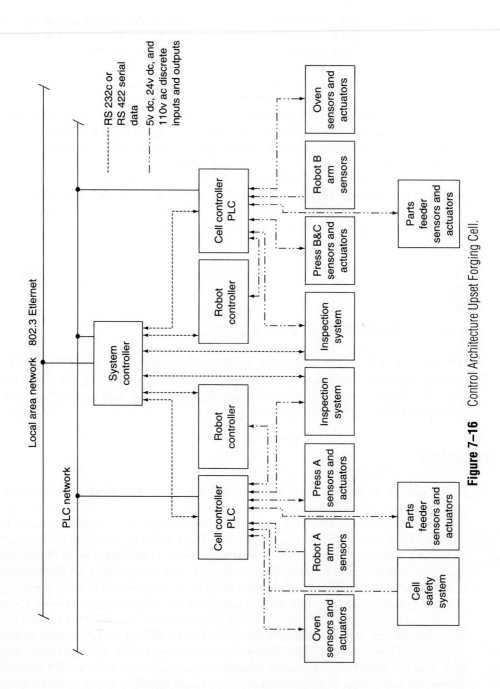

Figure 7-16 Control Architecture Upset Forging Cell.

Local area network   802.3 Etlernet

PLC network

RS 232c or
RS 422 serial
data

5v dc, 24v dc, and
110v ac discrete
inputs and outputs

System
controller

Robot
controller

Cell controller
PLC

Cell controller
PLC

Robot
controller

Cell controller
PLC

Oven
sensors and
actuators

Robot B
arm
sensors

Press B&C
sensors and
actuators

Inspection
system

Inspection
system

Press A
sensors and
actuators

Robot A
arm
sensors

Oven
sensors and
actuators

Parts
feeder
sensors and
actuators

Parts
feeder
sensors and
actuators

Cell safety
system

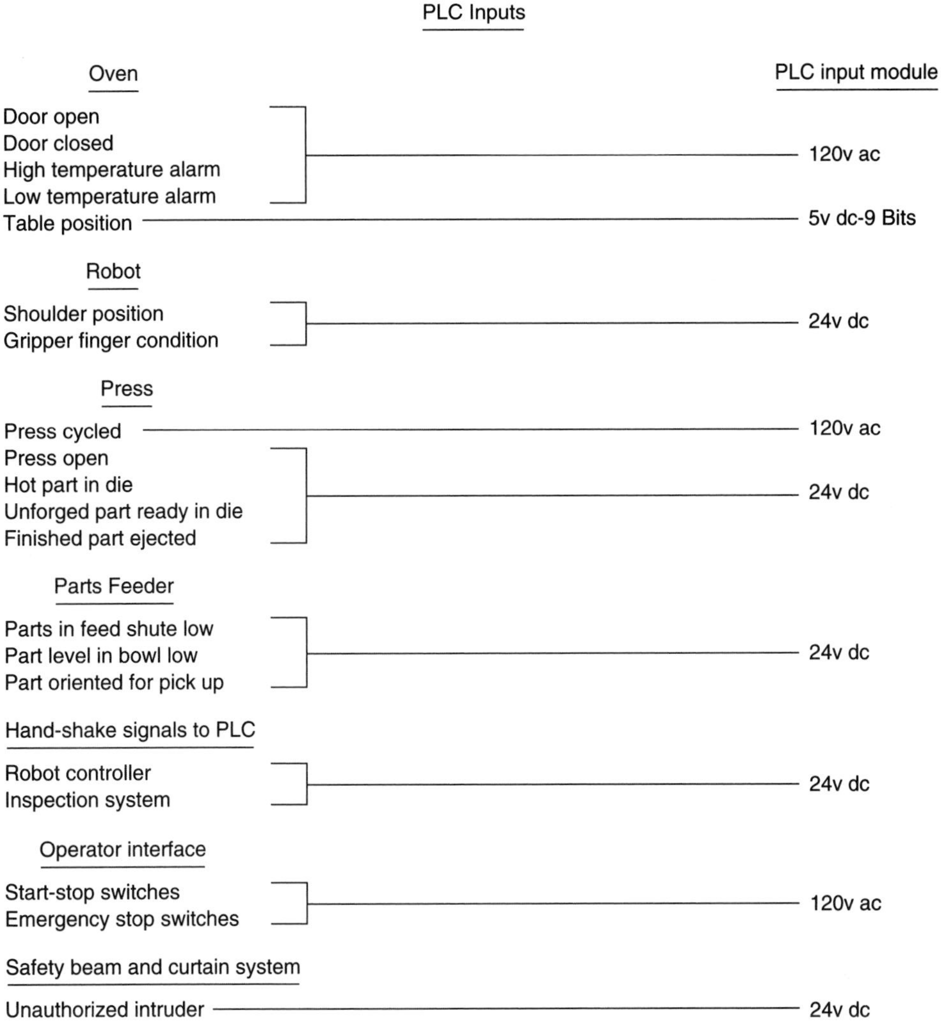

Oven   PLC input module

Door open
Door closed
High temperature alarm   — 120v ac
Low temperature alarm
Table position   — 5v dc-9 Bits

Robot

Shoulder position   — 24v dc
Gripper finger condition

Press

Press cycled   — 120v ac
Press open
Hot part in die   — 24v dc
Unforged part ready in die
Finished part ejected

Parts Feeder

Parts in feed shute low
Part level in bowl low   — 24v dc
Part oriented for pick up

Hand-shake signals to PLC

Robot controller   — 24v dc
Inspection system

Operator interface

Start-stop switches   — 120v ac
Emergency stop switches

Safety beam and curtain system

Unauthorized intruder   — 24v dc

Figure 7–17   Programmable Logic Controller Interface Drawing for Upset
Forging Cell.

The robot interface permits the controller to communicate with all the systems in the work cell. These systems range from passive switches with no on-board intelligence to complex vision systems with dedicated minicomputers. The controller interface must be compatible with this broad range of signal requirements. All robot interfaces fall into one of the following categories: simple sensor interface, complex sensor interface, and enterprise data interface. The simple sensor interface handles discrete signal sensors that do not require signal conditioning. The complex sensor interface provides signal conditioning for complex sensory signals before they are routed to a higher level controller. Finally, the enterprise data interface establishes the data connectivity between the robot and computer nodes in the enterprise network system.

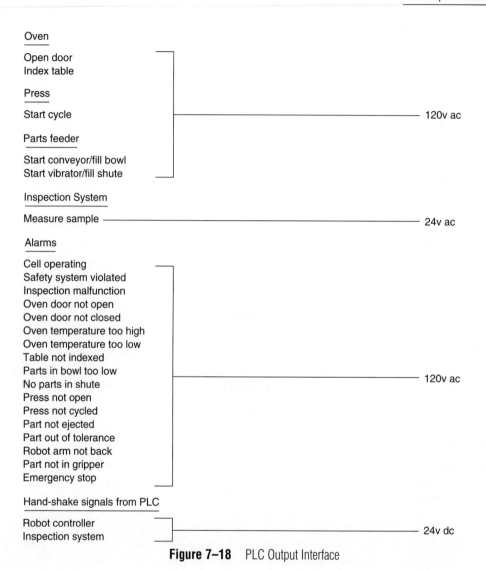

PLC Outputs

PLC output module

Oven

Open door
Index table

Press

Start cycle — 120v ac

Parts feeder

Start conveyor/fill bowl
Start vibrator/fill shute

Inspection System

Measure sample — 24v ac

Alarms

Cell operating
Safety system violated
Inspection malfunction
Oven door not open
Oven door not closed
Oven temperature too high
Oven temperature too low
Table not indexed
Parts in bowl too low
No parts in shute — 120v ac
Press not open
Press not cycled
Part not ejected
Part out of tolerance
Robot arm not back
Part not in gripper
Emergency stop

Hand-shake signals from PLC

Robot controller
Inspection system — 24v dc

Figure 7–18   PLC Output Interface

QUESTIONS

1. Describe the function of the types of software present in cell controllers.
2. Define the term *multitasking* and give three examples of software that supports that type of operation.

3. Compare the strengths and weaknesses of the three work-cell management software options.
4. List the components of a PLC system and describe the function of each.
5. Compare and contrast the robot controller block diagram from Chapter 1 with the block diagram for a CNC machine presented in this chapter.
6. Compare the function and operation of the controller used with servo and nonservo robots and describe the differences and similarities.
7. Compare the programming, control techniques, and operation of the servo and nonservo cell control.
8. Compare the simple and complex sensor interface.
9. What is the function of the cell controllers in Figure 7–15?
10. List the devices that would use a simple sensor interface and those that would be considered complex sensor interfaces in Figure 7–15.
11. How do each of the work cells in Figure 7–15 verify that the programs in the production machines are correct for the stock and parts delivered by the material-handling system?
12. Why are the work-cell sensors in Figure 7–15 interfaced to the PLC and not the robot controller?
13. Describe the difference between the ARTIC interface and communications controller in the work-cell controllers in Figure 7–15.
14. How do the functions of work-cell controllers and PLCs differ?
15. What are the characteristics of sensors using the complex sensor interface?
16. Describe the memory systems used in the robot controllers.
17. Name the two different types of interfaces handled by the robot controller CPU?
18. Describe one example of each type of interface named in number 17.

CASE PROJECTS

1. Use the W-E cell architecture as an example and develop a work-cell architecture for the extrusion work cell developed in question 2 of the case/projects section in Chapter 4.
2. Use the W-E PLC interface as an example and draw the wiring interface between the PLC, robot controller, sensors, operator interface, and other equipment in the extrusion work cell developed in question 2 of the case/projects section in Chapter 4.
3. What advantages did the W-E team hope to gain by using a separate PLC for each half of the production cell?
4. What impact would the use of a single PLC have on the wiring, programming, and operation of the upset forging cell.
5. What advantages did the W-E team hope to gain by using one cell controller for both sides of the process?

Work-Cell Programming

8-1 INTRODUCTION

Programming languages are the basic communication mechanisms between human beings and intelligent machines. Initially, these intelligent machines were the computers themselves, programmed to solve problems in business or scientific areas. More recently, however, computers have been incorporated into other industrial and office machines to increase their efficiency and capability. Despite this rapid integration of computers into machines of all types, the function of the programming language remains unchanged. Communication continues to be the primary function, but the ever-increasing number of communication sources requires an ever-increasing variety of data and information.

The first robots and automated cells were driven by drum-type sequencers (Figure 7–7) and provided little programming flexibility. Present-day automation is controlled by powerful digital computers, many with multiple processors, which permit a high level of user and machine communication. These computers are responsible for the following system functions:

- *Manipulation:* The control of the motion of all robot joints and work-cell actuators. This includes position, velocity, and path control of the robot arm during all programmed motion and sequencing the motion of drive devices such as motors and pneumatic actuators.
- *Sensing:* The gathering of information from the physical work cell. This includes the collection of sensory information and the control of peripheral equipment.

- *Intelligence:* The ability to use information gathered from the work cell to modify system operation or to select various preprogrammed paths and routines.
- *Data processing:* The capability to use databases and to communicate with other intelligent machines. This includes the capability to keep records, exchange programs, generate reports, and control activity in the work cell.

A manufacturing system with these characteristics would have production flexibility not found in most current small-batch systems. If these characteristics are to be realized in small-batch systems, however, the users must be able to specify how a system must operate to produce the product or to perform the task required. Consequently, the need for a suitable programming language that satisfies the system requirements is apparent. This chapter analyzes the current languages used by robots and other cell control devices.

8-2 WORK-CELL CONTROLLER PROGRAMMING

Work-cell software falls into one of three categories: in-house developed systems, application enablers, and open system interconnect (OSI) such as the *manufacturing message specification* (MMS) ISO standard 9506. The first two categories refer to proprietary software systems, because they are either written for a specific application or are developed around a third-party software management shell, called an *application enabler.* The third category is a standard information exchange software language supported by many of the vendors supplying automation equipment. Application enabler software, however, is used in most automation work cells to program the flow of information through the cell controller.

Let us consider an example application to clarify the three programming techniques. Review the operation of the machining cell in Figure 7–15 and note that the cell controller must pass, or download, a CNC program to the mill when the material for a new or different part is identified by the bar code.

In-House–Developed Software

As the name implies, in-house software is written by the end user using a programming language such as C. Programs developed in-house are written to address the specific control needs of the cell, and the work-cell control program is well integrated with other in-house–developed software for other enterprise areas. For example, the in-house–developed program in the cell controller (Figure 7–15) polls, or electronically interrogates, the bar code concentrator to determine if a new material code was read. If a new code is present, the software in the controller reads the new code, finds the correct CNC program for the part present, and downloads the code to the mill. This data exchange assumes a compatible electrical interface between the devices and that communication software is available in the controller to talk to the concentrator and mill. In-house program development provides the best opportunity for integration of information and data with other enterprise software solutions. In addition, machines that need custom communication drivers present no problems because the expertise is present to do the programming. Nevertheless, these advantages are offset by major disadvantages in development time and cost and the inability to easily change the software when the cell hardware or configuration changes.

Enabler Software

Common impediments to the development of a fully implemented CIM work cell are the exorbitant cost, time, complexity, and inflexibility of custom-programmed CIM solutions. The introduction of enabler software shifts the development of application programs from software engineers to manufacturing engineers. This shift occurs because enabler software provides a set of software productivity tools for the development of control software for the CIM cells. Products such as *Plantworks* from IBM, *Industrial Precision Tool Kit* from Hewlett-Packard, *CELLworks* from FASTech, and *Factory Link* from U.S. Data help reduce the difficulty in developing cell control and management applications.

These enablers have a library of driver programs for many commonly used production machines and machine controllers. In addition, they offer LAN and serial data communications support, mathematics and logic functions for internal computations, links to commonly used mainframe and microcomputer relational databases, real-time data logging, real-time generation of graphics and animation of cell processes, alarm and event supervision, statistical process control, batch recipe functions, timed events and intervals, counting functions, and the ability to write custom applications.

Advantages of enablers are an estimated tenfold improvement in cell control and ease of program development. The primary disadvantage is that the cell control is tied to a third-party software solution, so selecting the software that closely matches the application is critical. In addition, an enabler is written to satisfy the average process requirement; therefore, if a specific requirement is not covered, then additional programming inside the enabler is necessary.

OSI Solution

The most frequently used open system interconnect solution for cell control is the manufacturing message specification. The MMS is a standard (ISO 9506) for network communication between intelligent devices in a production environment. The standard has three parts: service specifications, protocol specifications, and robot interface and protocol specifications. The MMS standard defines a set of objects that exist within a device; for example, the MMS object could be the position of one axis of the robot arm. In addition, the MMS defines a set of communication services to access and manipulate the objects and describes how the devices will respond.

Implementing an MMS solution requires that all the devices support the MMS protocol and that they be linked by a manufacturing automation protocol (MAP) ethernet or broadband manufacturing network. The work-cell system in Figure 7–15 is reconfigured as an MMS solution in Figure 8–1. Take a few minutes and compare the two implementations.

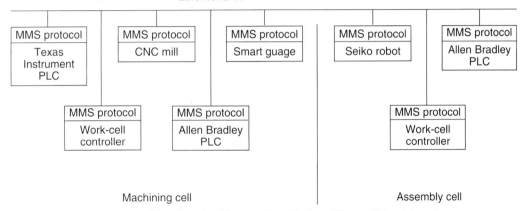

Figure 8–1 Manufacturing Message Specification. (*Source:* Rehg James A., *Computer Integrated Manufacturing.* Englewood Cliffs, NJ: Prentice Hall; 1994, p. 382)

Note in Figure 8–1 that all of the computer-controlled devices support the MMS and are on an ethernet-driven network. Also, there is no longer a need for the proprietary PLC network because the PLCs share data directly with every machine over the single LAN. The interface to the sensors would not change from that shown in Figure 7–15. With the MMS, the work-cell controller is no longer a data concentrator because MMS machines exchange data directly over the LAN. In the MMS configuration, each smart device provides real-time data without the need for work-cell management software to collect cell information. Many computer, robot, and PLC vendors have MMS support available as an option.

An advantage of the MMS is that it is a common communication standard, not a third-party vendor solution. Information is requested directly from the target device and delivered directly to the end user; no intermediate computer and software system are required. The primary disadvantages currently are the small user base and the limited number of equipment vendors who have agreed to support the standard.

8-3 PROGRAMMING SEQUENTIAL CELL ACTIVITY

The software described in the previous section provides programmed control of data and information exchange between the computer controller devices in the work cell. At a lower level in the cell control architecture, the action and movement of individual devices must be controlled. For example, when a sensor is activated by the placement of a part into the milling fixture in Figure 7–15, an orderly sequence of events must occur. The part clamping devices must be activated and a start-cycle signal must be sent to the mill. In most implementations the signals are discrete with voltages ranging from 5 volts dc to 110 volts ac. In most applications, sequential control is performed by PLCs; in some cases, however, the robot controller provides this control function.

PLC Programming

The technique used to illustrate the logic in PLC programs has changed little since the introduction of the first programmable logic controller in the 1970s. The process, called *ladder logic programming,* is a variation of the *two-wire diagrams* used to document the wiring of industrial control circuits. For example, the necessary switches, sensors, and controls to sequence the operation of a pump are drawn between the two vertical lines (L1 and L2) in the two-wire diagram in Figure 8–2. S1 and S2 are switches, C is a relay coil with contacts A1 and A2, M is a motor starter, and a valve is identified. Study the circuit and determine the operational logic present.

The circuit works as follows: (1) switch S1 is closed manually and causes the relay coil C to be energized; (2) when the relay is energized, the contacts A1 and A2 close; (3) the closing of contact A1 energizes the input valve and allows liquid to flow into the tank; (4) the closing of contact A2 causes no immediate action; (5) switch S2 is closed manually and causes the pump to operate. Note that the change in contact A2 in step 4 was necessary for pump operation. If the valve switch S1 is manually opened while the pump is operating, the change in contact A2 opens up the circuit which is energizing the pump, and it stops. Review this operation and study the two-wire diagram in Figure 8–2 until the operation of this circuit is clear. Also, verify that you understand the logic conditions required for the valve and pump to operate.

The PLC ladder logic program that would provide the same logical control as the circuit in Figure 8–2 is illustrated in Figure 8–3a. The PLC processor *scans* all the inputs and writes the *on* or *off* state of each input into a memory location. The input conditions are applied to the programmed logic in the processor memory, the appropriate output conditions are determined, and the output terminals are set either *high* or *low* accordingly. The example program in Figure 8–3a illustrates how the processor uses input conditions to change output states. Study the ladder logic program (Figure 8–3a) and the PLC wiring interface (Figure 8–3b).

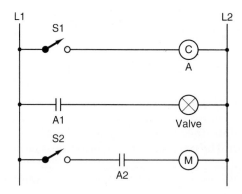

Figure 8–2 Two-Wire Diagram of Pump System. (*Source:* Rehg, James A., *Computer Integrated Manufacturing.* Englewood Cliffs, NJ: Prentice Hall; 1994, p. 389)

The symbol identified as I1 in ladder rung 01 represents the logical control provided by input 1 on the input module. If a voltage is present at input 1 (the valve control switch S1 is closed) then I1 closes and turns *on* CR1. With no voltage at input 1, I1 is open and CR1 is *off*. CR1, called a *control relay*, is created in the PLC software operating system in the processor for logical control requirements. In most PLCs the processor permits the programmers to use large numbers of control relays for logical control needs.

When CR1 is turned to the logical on state, all the contacts labeled CR1 change states. When the CR1 contact in rung 02 closes, the PLC processor causes the output module terminal 1 to switch to a voltage state, which causes the solenoid valve to energize. This action causes liquid to enter the tank.

When the pump motor control switch S2 closes, the input 2 of the input module has voltage applied, and the I2 contacts close. The two contacts in rung 03, CR1 and I2, are now both closed so the O2 output is energized. With O2 active, output

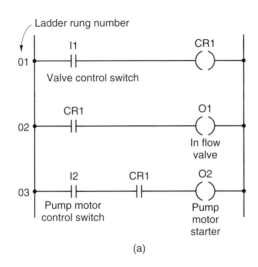

(a)

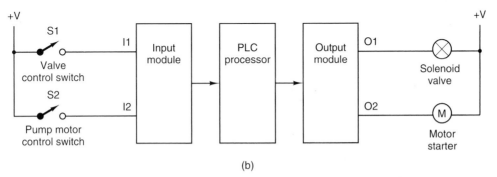

(b)

Figure 8–3 PLC Program and Input/Output Interface: (a) Ladder Logic Program and (b) PLC Wiring Interface. (*Source:* Rehg, James A., *Computer Integrated Manufacturing.* Englewood Cliffs, NJ: Prentice Hall; 1994, p. 390)

2 is low, the motor starter is energized, and the pump motor is running. The pump motor cannot be turned *on* if the valve switch is open, because the contacts CR1 would be open. The logical operations and control performed by PLC processors include *AND, OR, INVERSION, timers, counters,* and *comparators.*

Other Sequential Programming Options

Although ladder logic offers a familiar programming environment for industrial applications, the process has some disadvantages. For example, the unstructured nature of the process results in multiple solutions for the same control problem. In addition, interlocks are required to eliminate the undesired interaction between two different parts of the ladder logic program. Also, large ladder logic programs are difficult to troubleshoot because the contacts of a control relay could appear anywhere on the ladder. PLC vendors have developed a variety of solutions to these problems.

A European standard, called *Grafcet,* is used to overcome many of the problems associated with conventional ladder logic programming. Grafcet is a graphical method of functional analysis that represents the functions of sequentially automated machines as a sequence of steps and transitions. Each step includes commands for control of the machine or system which are either active or inactive. If a step is active, then the commands present in the step are executed. The flow of control in Grafcet passes from one step to another through conditional transitions that are either *true* or *false.* If the transition is true, then control passes from the current step to the next. After the transition, the previous step is inactive and the current step is active. In a sequential process, each control function is represented by a group of steps and transitions called a *function chart.* A number of companies have adopted programming standards that emulate the Grafcet standard; for example, Allen Bradley has *sequential function charts* and TI/Siemens has *stage programming.*

8-4 ROBOT LANGUAGE DEVELOPMENT

Initially, robot languages were designed using two techniques. The first approach focused on developing a language that satisfied the control needs of the robot arm. Next the language was expanded to include language structures, for example, conditional branching and input/output interfacing. The T3 language developed by Cincinnati Milacron for its industrial robot family was an example of this type. Good control of the robot's manipulation and tool path resulted from this approach, but the technique did not fit the structure normally associated with computer languages and did not support the data processing function well.

The second technique started with an existing general-purpose computer language such as BASIC or FORTRAN and added the robot control commands. VAL, a language developed for the early Unimation PUMA robots, used BASIC as a base and added motion control commands. This technique produced a robot language that was well defined, operational, and easy for programmers to use. The use of a fully functional computer language, however, forced some design compromises that made motion control commands less efficient.

Currently, robot language design takes a top-down approach that can produce a new general-purpose language capable of supporting all four system functions: manipulation, sensing, intelligence, and data processing. AML (A Manufacturing Language) from IBM is an example of this new approach.

Few standards currently exist for robot control languages. There is no interchangeability of programs among manufacturers—and in some cases, only limited interchangeability of programs between models from the same manufacturer.

8-5 LANGUAGE CLASSIFICATION

One way to classify the many languages used by robot manufacturers is according to the level at which the user must interact with the system during the programming process. For example, if the user must specify the joint angles for each move, the level of interaction is very low compared with a language that permits the user to specify the motion required in statements such as "Pick up the part." Using this criterion, we can group the current languages into four loosely formulated levels. Of course, overlaps between levels exist and some languages appear to straddle two levels, but the classification process is still valuable. Figure 8–4 shows the four basic levels into which all robot languages are grouped.

Joint-Control Languages

Languages at this level concentrate on the physical control of robot motion in terms of joints or axes. The program commands must include the required angular change of rotational joints or the extension lengths of linear actuators. The language usually does not support system or work-cell commands, such as INPUT or OUTPUT, which can be incorporated into the programs of higher-level robot languages for control of external devices.

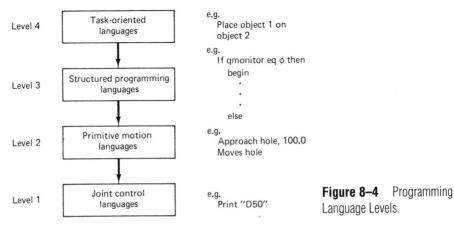

Figure 8–4 Programming Language Levels.

In this level language, the user must program in joint space. The term *joint space* means that all the programmed points in the work envelope of the robot are expressed as a series of axis positions for all the axes on the arm. Figure 8–5 illustrates a robot arm that has three programmed points. The table included in the figure shows the joint angles that are required for the robot axes for each programmed point.

This level of robot language is used on some less sophisticated *point-to-point* servo machines and on all *stop-to-stop* pneumatic robots controlled with PLCs. The language used for stop-to-stop robots (eg, the robot in Figure 2–11) is the general-purpose language of the PLC.

Primitive Motion Languages

Point-to-point primitive motion languages are now usually confined to older robot programming languages, or they may be an optional programming mode for a more sophisticated robot language. Although the languages included in this group vary widely, they all exhibit the following characteristics:

- A program point is generated by moving the robot to a desired point and depressing a program switch. A sequence of points is saved in this manner, thereby producing a complete program.
- Program editing capability is provided.
- Teaching motion of the robot is controlled by either a teach pendant, terminal, or joystick.
- The programmed and teaching motion can occur in the Cartesian, cylindrical, or hand coordinate modes.
- Interfacing to work-cell equipment is possible. Robot controllers can interact with external signals by using the external signals for control or by signaling external events. Work-cell control by the robot system is possible.
- The language permits simple subroutines and branching.

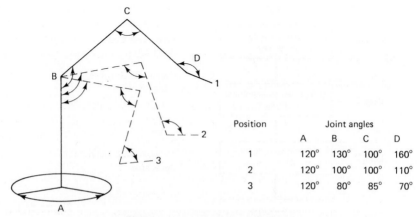

Position	Joint angles			
	A	B	C	D
1	120°	130°	100°	160°
2	120°	100°	100°	110°
3	120°	80°	85°	70°

Figure 8–5   Joint-Space Programming.

In addition, some of these point-to-point languages permit simple parallel execution using two or more arms in the same work space. Some languages in this group have limited coordinate-transformation capabilities. The VAL language, for example, can define reference frames, invert transformations, and multiply matrices.

The primary advantage of languages at this level is proven performance on the manufacturing floor. The common disadvantage is that the emphasis in programming is still on robot motion rather than on the production problem. In addition, this level does not support the need for *off-line* programming.

Structured Programming Languages

The structured programming languages offer a major improvement over the primitive motion level and have become the standard for the major vendors of robots. This level is composed of languages that have the following characteristics:

- A structured control format is present.
- Extensive use of coordinate transformations and reference frames is permitted.
- Complex data structures are supported.
- Improved sensor commands and parallel processing above the previous language level are included.
- System variables (called state variables), whose value is a function of the state or position of the system, are permitted.
- The format encourages extensive use of branching and subroutines defined by the user.
- Communication capability with local area networks is improved.
- *Off-line* programming is supported.

Not all of the languages in this classification exhibit all of the characteristics listed, but each of them has features that make it appropriate for the structured classification group.

The primary advantage of using a language at this level is the programming advantage gained from the use of transformations. This becomes especially true in complex assembly applications and in support of off-line programming efforts. The major drawback at this level is the increased educational demand placed on the user, who must program with transformations in a structured format.

Task-Oriented Languages

The primary function of a task-oriented language is to conceal from the user the commands and program structure that normally must be written by the programmer. The user need only be concerned with solving the manufacturing problem. Languages at this level have the following characteristics:

- Programming in natural language is permitted. A natural language command might be "Put bracket A on top of bracket B."
- A plan generation feature allows replanning of robot motion to avoid undesirable situations.

■ A world modeling system permits the robot to keep track of objects. This feature provides the systems needed to locate and identify objects, to determine pickup point and orientation of objects, and to move objects relative to one another. In addition, the system can store and use the new relationship between two joined objects.

■ The inclusion of collision avoidance permits accident-free motion.

■ Teaching can be accomplished by showing the robot an example of a solution.

Currently, no languages at this level are operational, but significant efforts are underway at university and industrial research laboratories. Several experimental languages have exhibited the characteristics of this level: AUTOPASS (Automatic Programming System for Mechanical Assembly) from IBM, RAPT (Robot Automatically Programmed Tools), and LAMA (Language for Automatic Mechanical Assembly).

A review of industrial applications indicates that level 3 structured programming languages are now the most frequently used in robot work cells. Figure 8–6 provides a partial list of languages and the corresponding level of their operation.

8-6 ROBOT PROGRAM FUNDAMENTALS

A robot program is a set of instructions that causes the system to perform the desired task. The program has two basic parts.

The first part is a set of points in space, called *translation points* or *position points*, to which the robot will move when the program is executed. The number

Origin	Level 2	Level 3	Level 4
ABB		RAPID	
Adept		V	
		V+	
Cincinnati Milacron	T3		
GMFanuc		KARL	
IBM		AML	AUTOPASS
		AML/E	
Kawasaki		AS	
McDonnell Douglas		MCL	
Panasonic		PARL-I	
Rhino	RoboTalk		
Sankyo		Sankyo language	
Seiko		DARL II	
Unimation	VAL	VAL II	

Figure 8–6 Programming Languages by Level.

and location of translation points for servo and nonservo robots are quite different. Any location in the work envelope of a servo robot can be programmed as a position or translation point. For example, the Seiko robot in Figure 8–7 has a cylindrical work envelope and the three translation points (T5, T10, and T20) identified in Figure 8–8. However, the number of points that could be programmed is only limited by the resolution of the servo system driving the arm.

In contrast, the nonservo robot has a limited number of positions, which are determined by the number of fixed and programmable stops. The nonservo robot

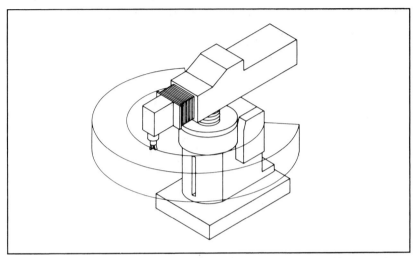

Figure 8–7   Seiko Robot Using DARL Programming Language. (Courtesy of Seiko Instruments USA Inc.)

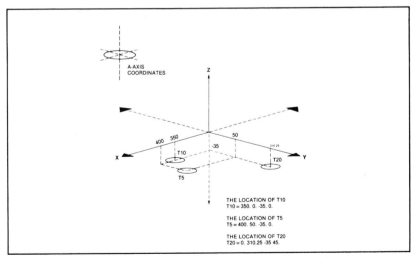

Figure 8–8   Programming Coordinate System for Seiko Robot. (Courtesy of Seiko Instruments USA Inc.)

system in Figure 8–9, for example, is a three-axes arm with one programmable intermediate stop and a wrist-roll axis. The standard arm, without the intermediate stop actuator, would have 8 programmable positions; with the intermediate stop added, the total number of positions is 12. Find the 12 positions for the robot arm in Figure 8–9.

The second part of a robot program consists of program statements that determine the order of motion, define the logic and conditions for decision, gather information on the operation of the cell, and analyze and prepare data for communication with other machines. Most servo robot languages have a robust set of command statements available to the programmer for control of the arm. The statements are similar to the structured commands used in computer languages such as Qbasic and Pascal. The statements are generally grouped into a number of categories based on the function performed by the command. In the PARL-I language from Panasonic, for example, the control functions include movement, input/output, external signal, communications, time, form, and program.

Nonservo robots are normally controlled by sequencing the pneumatic or hydraulic valves with a PLC. Therefore, the programming language is ladder logic or function charts, and the command statements are dictated by the PLC chosen for robot arm control.

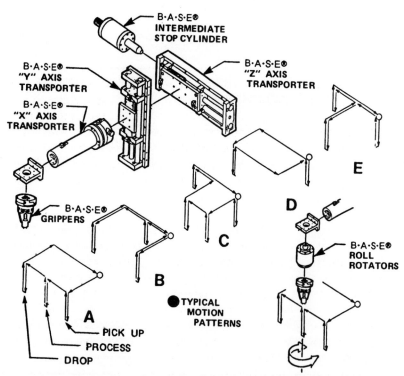

Figure 8–9   Nonservo Position Points. (Courtesy of Mack Corporation)

Each translation point contains a value for each axis in the robot coordinate system. For example, the four-axes Seiko robot and work envelope in Figure 8–7 requires four values (X, Y, Z, and A) for every translation point. Three translation points (T5, T10, and T20) are illustrated in the Cartesian coordinate system pictured in Figure 8–8. For example, the T5 translation point is located at X = 400 mm, V = 50 mm, Z = − 35 mm, and the A rotation of the gripper is 0 degrees. In stop-to-stop or nonservo robots, the translation points are defined by the fixed or hard stops.

8-7 TRANSLATION OR POSITION POINTS FOR SERVO ROBOTS

The translation points are locations inside the work envelope of the robot and represent positions through which the tool center point will pass. The translations are always defined in relation to the origin of a Cartesian coordinate system located in a frame of reference. In the example illustrated in Figure 8–8 the first three values in each data point (see bottom right of figure) represent the distance in millimeters from the origin of the Seiko reference frame to the translation point; the last value is the rotation of the gripper in degrees. The reference frame system used in programming is a natural place to begin our detailed discussion of translation points.

Reference Frames

Reference frames are sets of Cartesian coordinates that describe the positional relationship between the robot, the gripper, the tool center point, the workpiece, and the universe (world) in which all of these exist. The dimensional relationships between the parts of the robot and the production parts in the work cell are described by means of mathematical equations called transformations. Figure 8–10 illustrates a robot reference frame that is located in a universe frame. In addition, the reference frames for the workpiece, workpiece holder, and gripper or tool center point are shown. The robot controller can calculate the necessary joint angles that will align the tool center reference frame with the part or workpiece reference frame.

The program stored in the controller performs the calculations using the transformation equations written for the type of robot arm used. Although the transformation equations are completely transparent to the robot user, the programming languages use the reference frame as a programming tool. For example, the DARL language from Seiko has a command, **DEF FR** (define frame), that permits an entire sequence of programmed points to be changed by changing the reference frame. A move command in DARL, such as **MOVE T10,** can produce different motions since the point identifier T10 can have different coordinate values in each reference frame. Reference frames are also used in work-cell simulation systems to determine the optimum location of all the equipment in a production work cell that uses a robot. The simulation application note in this chapter

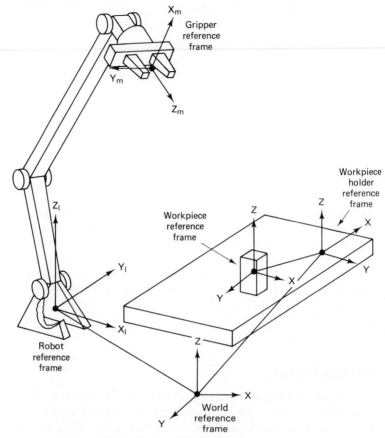

Figure 8–10   Reference Frames in Work Cell.

describes a system developed by the McDonnell Douglas Corporation. Although a complete description of reference frames and of transformation equations is not appropriate for this text, a basic understanding is essential for the robot technician, the applications programmer, and the work-cell system designer.

The position, or angle, of every servo-driven robot joint is measured by sensors and the information is fed back to the controller. Based on the position information, the location of the gripper with respect to the robot frame is calculated by the robot controller. For example, the X, Y, and Z values for the gripper in the robot reference frame in Figure 8–10 are calculated using equations that define the relationship between the gripper reference frame and the robot reference frame. The equations have variables for each joint angle of the robot arm and constants, such as the length of each arm element and the distance from the tool plate to the tool center point. As the joint angles change, the displacement of the gripper from the X, Y, and Z axes of the robot is calculated. Similarly, the equations that relate the robot reference frame to the universe reference frame determine the position and orientation of the gripper in the universe reference frame. Once the relationship

Complex manufacturing work cells represent a considerable capital investment, and the electrical and mechanical requirements of all the systems in the cell must be considered for an effective design. The addition of a robot into a complex manufacturing work cell creates many new and unique problems. For example, the location of the robot in the cell is a function of the robot's geometry, payload capacity, end-of-arm tooling, and speed and the location of other cell hardware. Positioning the robot is a three-dimensional problem because the robot has three-dimensional movement. Work-cell simulation systems solve this problem by graphically displaying the robot arm and work-cell hardware on a graphics computer monitor with full dynamic motion capability. The advantages of robot work-cell simu-

lation include the ability to (1) select the most efficient geometry and robot size for the given work-cell application; (2) simulate the actual size and motion of a wide range of currently available industrial robot arms; (3) construct a scaled three-dimensional computer model of the work cell; (4) test the model for interference and reach over the full range of robot arm movement; (5) run the work cell and robot over a range of production speeds to determine optimum cycle times; (6) generate the robot program required for the application and for the arm selected.

The figure below shows a printout of a simulation for an ABB robot palletizing parts in bins. Note the X, Y, Z reference frames for each part in the work cell. The following simulation software is frequently used to program

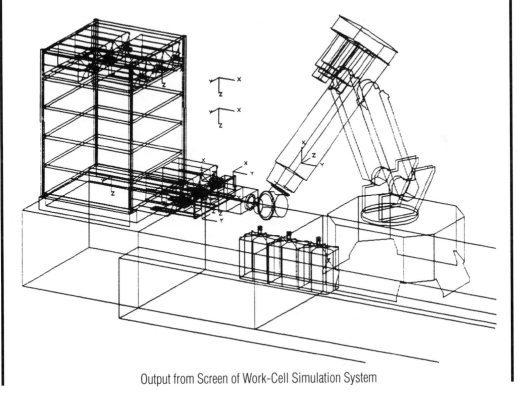

Output from Screen of Work-Cell Simulation System

and simulate robot and work-cell environments: WORKSPACE, Robcad, IGRIP, and Grasp. These tools, called computer-aided software engineering (CASE) software, enhance and shorten the work-cell design and robot programming process. The features supported by these packages are three-dimensional CAD editor, three-dimensional solid shading, off-line programming language translators, intertask communications, cycle time calculations, collision detection, reach analysis, integrated robot program editor, robot calibration, work-cell calibration, AutoCAD .dxf file import capability, solid modeling with constructive solid geometry features, dimensioning, torque analysis and motion simulation, user definable kinematics, modeling of parallelograms, auxiliary robot axes, linking of user-defined programs, automatic inverse kinematic solution generator, computer-aided learning (CAL), and support for a microcomputer platform.

between the reference frames is defined by tranformation equations, the displacement of the part and the gripper from the universe reference frame is determined by the controller. Thus, for the robot to grasp the part, the controller only needs to change the joint angles until the gripper reference frame and part reference frame have the same displacement and orientation from the universe reference frame. Under these conditions (Figure 8–11), the reference frame of the gripper and the reference frame of the part are aligned. This alignment would not have been possible without the equations that describe the relationship between the reference frames in the system.

Reference frames for the robot and work-cell environment are required for some programming operations. For example, the initial reference frame for the Seiko robot is called *Frame 0* or the *HOME* frame. Nine additional reference frames can be defined and the relationships between the frames can be established. The reference frames identified in Figure 8–10 (gripper, workpiece, workpiece holder, and world) could be assigned to the nine available in the Seiko system.

Programming Servo Robot Translation Points

Five basic methods are used to define translation points: (1) teaching of translations by robot positioning; (2) teaching of translations by coordinate input; (3) creation of new translations by algebraic expressions; (4) creation of a new translation by means of an external interrupt; and (5) downloading new translations through the RS-232C port.

Method 1—Teaching Points by Robot Positioning. This type of programming is available at all language levels and remains the most frequently used technique for all types of robots. With the controller in the manual or monitor mode, the tooling is visually moved into the correct position using the manual teach pendant; then the translation point is programmed. On some robots, programming occurs

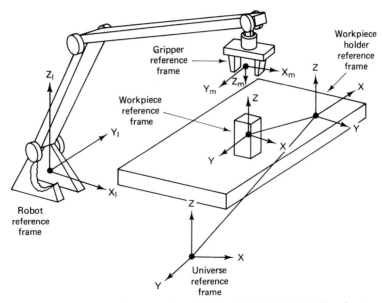

Figure 8–11 Reference Frames in Work Cell With Parts Aligned.

by just pressing a button on the teach pendant. On other systems, for example the Seiko, a command (**HERE T** <**0–399**>) is entered from the keyboard. The Seiko allows four hundred translation points (0 to 399) to be programmed.

Most robots have several different operational modes to assist in visually positioning the tooling when this method of programming is selected. The PUMA robots using the VAL language, for example, have four modes: *WORLD*, *TOOL*, *JOINT*, and *FREE*.

- *WORLD:* In the WORLD mode the coordinate system is fixed to the base of the robot. In this mode, pressing the X, Y, or Z direction buttons on the teach pendant causes linear movement along any of the three axes. For example, pressing the +X motion button on the PUMA teach pendant causes the tooling to move along the X axis in the positive direction. This mode is used most often to teach points visually.

- *TOOL:* In the TOOL mode the coordinate system is fixed to end-of-arm tooling with the Z axis coincident with the longitudinal axis of the gripper. Pressing the Z motion buttons causes the tooling to move along its longitudinal axis. As the tooling orientation is changed, the coordinate system follows the tooling. This is ideal for insertion of pins into holes or for loading a fixture that is not normal to any WORLD axis.

- *JOINT:* In the JOINT mode the movement of the tooling is not associated with any coordinate system. Each robot joint can be moved independently and is used for coarse moves in the work cell.

■ *FREE:* In the FREE mode the joints can be physically moved by manually overcoming the small holding torque present on each joint axes motor. This mode is used to physically move the arm away from a problem situation.

The four modes just described support visual programming by means of the teach pendant and are only used in method 1 for teaching translation points.

Method 2—Teaching Points by Coordinate Input.   Teaching by coordinate input requires that the position and orientation of the tool center point be entered as either a program statement or through the controller keyboard. The Seiko robot in Figures 8–7 and 8–8, for example, would require four values (X, Y, Z, and A), whereas the jointed-spherical arm in Figure 1–10 would require the six axes variables shown.

The following program statements demonstrate coordinate input programming on the Seiko robot:

$$T25 = 300.\ 350.\ -10.5\ 15.$$
$$MOVE\ T25$$

The first line creates the translation point T25 with values for X, Y, Z, and A; the second line directs the robot to move to the point just programmed. All level 3 languages permit this type of position programming.

Position points can also be programmed by direct input from the keyboard. On the Seiko robot the controller is put into the *Immediate Execution* mode, called *Monitor,* and the following command is entered from the keyboard:

$$DO\ T25 = 300.\ 350.\ -10.5\ 15.$$

The new translation point is created, and the robot moves to the new position. Level 2 and 3 languages support this type of procedure.

Method 3—Creating Points by Algebraic Expression.   New translation points can be created using previously defined translations in algebraic expressions. Translations can be added or subtracted and modified by an integer value. The following examples for the Seiko robot illustrate this operation:

Adding Translation Points

$$T10 = 20.\ 30.\ 5.\ 0.$$
$$T20 = 30.\ 20.\ -5.\ 50.$$
$$T30 = T10 + T20$$

The third line adds the corresponding coordinate values so the new translation point has a value of 50. 50. 0. 50.

Subtracting Translation Points

$$T10 = 20.\ 30.\ 5.\ 0.$$
$$T20 = 30.\ 20.\ -5.\ 50.$$
$$MOVE = T10 - T20$$

The third line subtracts the corresponding coordinate values and moves the robot to the new location at −10. 10. 10. −50.

Multiplying a Translation Point by an Integer

$$T10 = 20. 30. 5. 0.$$
$$T20 = 30. 20. -5. 50.$$
$$Y = 2$$
$$MOVE = T10*(-1) + T20*Y$$

(*Note:* * is a symbol for multiplication)

The fourth line adds the two modified translations (−20. −30. −5. 0.; 60. 40. −10. 100.) and moves to the new location (40. 10. −15. 100). Level 3 languages permit this type of position point modification.

Method 4—Creating Points by Use of External Interrupt. An external signal can be used to interrupt the current move command and trigger the creation of a new translation point. The new point is the position of the tool center point at the time of the interrupt. The DARL language from the Seiko robot is used to illustrate this command.

SEARCH +1E3 T10 THEN HERE T20
ELSE STOP

In this example the statement helps the robot find the top of a stack of a random number of plates (Figure 4–7). When this command is executed, the robot is moving a vacuum gripper and proximity sensor down on the top of a stack of plates (point T10 is at the bottom of the stack) that must be loaded into a machine. As the tooling moves toward T10 the controller continually scans the input 1E3, which is connected to the proximity sensor. The robot will continue moving vertically down until the top plate in the stack is sensed and input 1E3 goes *true*. The interrupt generated by 1E3 causes the *THEN* part of the statement to be executed and the location of the top plate becomes translation point T20. If translation point T10 is reached and no interrupt was generated, then no stack is present and the *ELSE* part of the statement is executed and the robot *stops*.

Creation of position points through interrupts is provided in most level 3 languages.

Method 5—Downloading Points through the RS-232C Port. This method permits translation or position points to be specified from an external source such as a host computer, cell controller, work-cell simulator, off-line programming station, or vision system. The data are transmitted in ASCII format over a serial communications protocol. Many level 3 languages support translation point data interchange over a serial link.

The creation of the translation points through which the tool center point will pass during the execution of the program is one part in the development of a manufacturing program for a robot.

Programming Nonservo Robot Translation Points

The translation or programmed points for nonservo robots are determined by the fixed stops built into the system. The motion of the stop-to-stop robot is controlled with a sequencer such as the drum controller in Figure 7–7 or by a programmable logic controller. The robot in Figure 8–9 has five different programmed sequences (A through E) illustrated. Note that HOME position is identified by the small circle in the path. The programmed motion is confined to the twelve positions defined by the end stops and the intermediate stop feature. The PLC interface for the robot is drawn in Figure 7–9.

8-8 PROGRAM STATEMENTS FOR SERVO ROBOTS

The second part of level 3 robot program languages is a list of *program statements* that will move the arm and tooling through the translation points described earlier and satisfy the production requirements. The steps in developing a robot program are (1) establish a basic program structure; (2) analyze the manufacturing process in which the robot will work; (3) divide the robot action into tasks and subtasks; (4) draw a *task point graph* that describes the desired motion; (5) identify the translation points on the task point graph; (6) assign values for all system variables to control the motion; (7) write and enter the command statements; (8) create or teach the translation points; and (9) test and debug the program. Several of these steps require clarification.

Basic Program Structure

The first step in developing a robot program is the establishment of the basic program structure (Figure 8–12). All programs start at the robot's *HOME* position

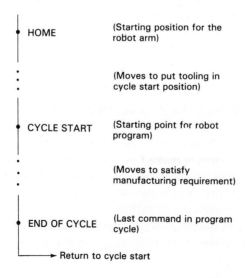

Figure 8–12 Basic Robot Program Structure.

and move out to a start point in the cycle called *CYCLE START*. The translation point for CYCLE START should be located as close as possible to the main motion path of the tooling because the robot returns to this point after every manufacturing cycle.

The final point in the program, called *END OF CYCLE*, provides a branching command to direct the program execution back to the CYCLE START point.

The program points that solve the manufacturing problem are located between the CYCLE START and END OF CYCLE points. The operations are divided into tasks and subtasks with the tasks embedded into the main body of the program and the subtasks included as subroutines called from the main body.

Process Analysis

The development of the program starts with an analysis of the problem that will be solved by the robot. The programmer must have a complete knowledge of the manufacturing process in which the robot functions. It is from this knowledge in the operation of the process that the programmer can identify the required motion and commands, divide the motion into tasks and subtasks, and establish the values of the system variables for each translation point.

Tasks and Subtasks

Based on a complete understanding of the production process, the programmer divides the required robot motion into tasks and subtasks. The following application of an injection molding machine illustrates this concept. A robot unloads a part from the injection molding machine, passes the part under a vision inspection system, and places good parts on an exit conveyor and bad parts in the recycling bin. Consider the following task breakdown. The major tasks include unloading the machine and submitting the part to the vision inspection system. Placing good parts on the conveyer and bad parts in the rework area would be subtasks. In this example the major tasks are in the main program; the subtasks are included as subroutines called from the main program. The following rules of thumb are used to separate tasks and subtasks:

- If alternative actions are selected based on input from external sources, then make each alternative path a subtask in a separate subroutine.
- If a particular robot motion or command sequence (eg, turning on an alarm to call an operator) is required in two or more places in the program, then make the action a subtask, and put it in a subroutine.
- If two or more sequences of major tasks are identified, then put each major task sequence in separate subroutines called from the main program. Assign subtasks for each major task to subroutines called from the major task subroutine.
- If a part of the robot's action is likely to require frequent modification or updating, then make the portion of the program a subtask, and place it in a subroutine.

After the robot action is divided into tasks and subtasks the structure of the total robot program is established.

Task Point Graph

A *task point graph* (TPG), step 4 in the programming steps, is a visual tool to illustrate the program flow and arm motion required for a manufacturing problem. The TPG includes all translation point data, the motion variables used at every point, and the logic used to make decisions during the program execution. A TPG for the robot and injection molding problem in the last section is drawn in Figure 8–13. Note that translation points are given descriptive names and that the subroutines are not shown.

On-Line and Off-Line Programming

The terms *on-line* and *off-line* programming define the location where the robot program is developed. For on-line programming, the production operation is stopped and the robot programmer puts the production machine into the programming mode. Then the programmer teaches the robot the required position, motion, and control sequences. The positions or translation points are taught by visually moving the production tooling to the exact work-cell location and entering the position into the program with a teach button on the teach pendant. In this method, the exact location of the work-cell components and the native accuracy of the robot system are not critical for good operation. The automation will work as long as the robot's repeatability is good and the location of work-cell machines and parts does not change. The major disadvantage in on-line programming is the lost production time.

Some robot languages use variable names for the translation points and permit the control structure, moves, and program logic to be developed on a word processor on a microcomputer or engineering workstation. The completed program is downloaded to the robot controller over a serial communication channel, and production is stopped only to teach translation points for all the location variables named in the program. A significant reduction in lost production results from this modified on-line programming technique. The VAL II program in Figure 8–14 illustrates the program structure that uses variable names (PALLET, CON) for translation points in the work cell. After this palletizing program is loaded into the controller, two translation points are taught and production begins.

The term *off-line programming* means that *all* of the programming is performed away from the robot and the production area. All translation points are calculated by the controller from coordinate values entered into the program in the off-line mode. For this technique to work, several conditions must exist:

■ The accuracy of the robot and the controller must be excellent and consistent for all the robots used in the production area.
■ All machines in the work cell must be accurately located relative to the robot reference frames.

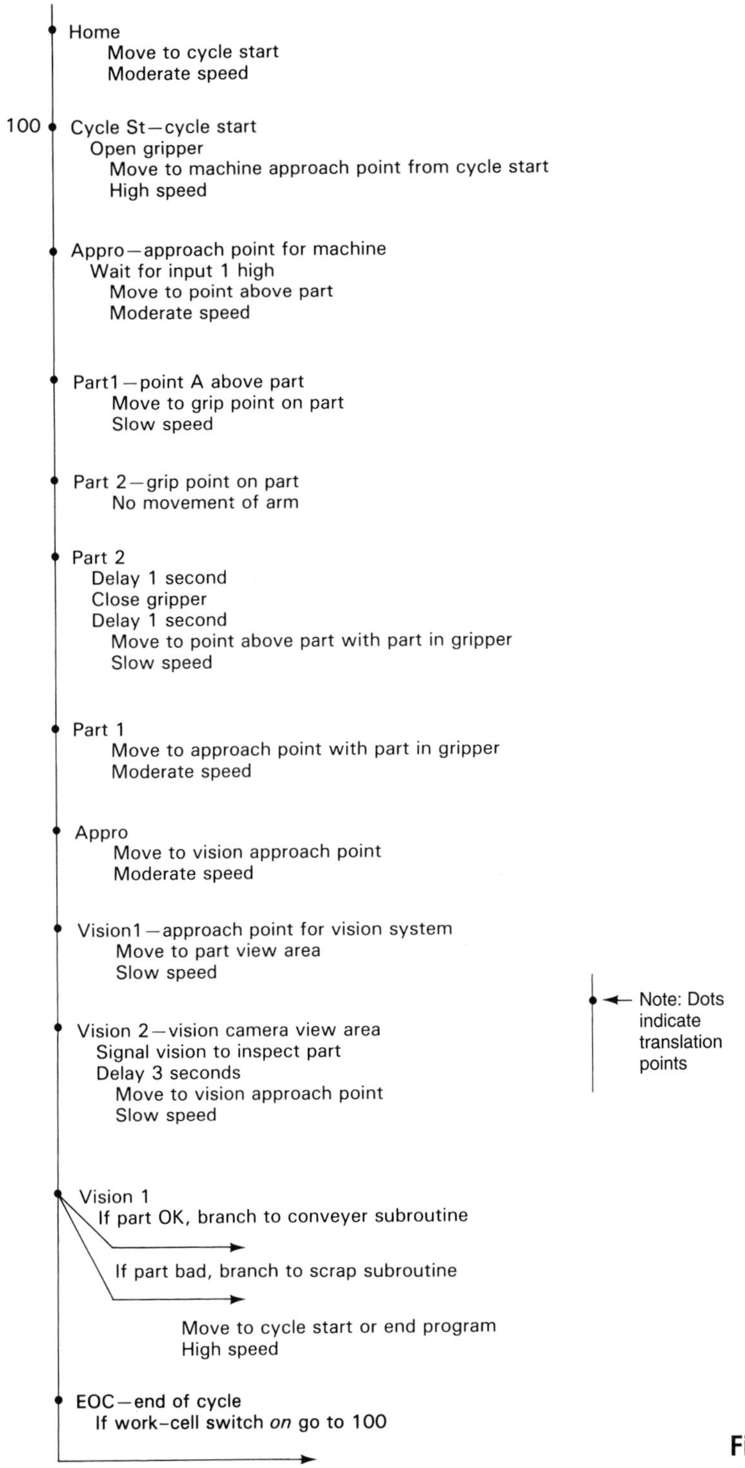

Home
    Move to cycle start
    Moderate speed

100   Cycle St—cycle start
    Open gripper
      Move to machine approach point from cycle start
    High speed

Appro—approach point for machine
    Wait for input 1 high
      Move to point above part
    Moderate speed

Part1—point A above part
      Move to grip point on part
    Slow speed

Part 2—grip point on part
      No movement of arm

Part 2
    Delay 1 second
    Close gripper
    Delay 1 second
      Move to point above part with part in gripper
    Slow speed

Part 1
      Move to approach point with part in gripper
    Moderate speed

Appro
      Move to vision approach point
    Moderate speed

Vision1—approach point for vision system
      Move to part view area
    Slow speed

Note: Dots indicate translation points

Vision 2—vision camera view area
    Signal vision to inspect part
    Delay 3 seconds
      Move to vision approach point
    Slow speed

Vision 1
If part OK, branch to conveyer subroutine

    If part bad, branch to scrap subroutine

      Move to cycle start or end program
      High speed

EOC—end of cycle
    If work-cell switch *on* go to 100

Figure 8–13   Task Point Graph.

```
1.          SETI      PX = 1
2.          SETI      PY = 1
3.    10    GOSUB     100
4.          IF PX = 3 THEN 20
5.          SHIFT PALLET BY 100.0,0,0
6.          GOTO 10
7.    20    IF PY = 3 THEN 40
8.          SETI PX = 1
9.          SETI PY = PY + 1
10.         SHIFT PALLET BY -900.0, 100.0,0
11.         GOTO 10
12.   100   APPRO CON,50
13.         WAIT CONRDY
14.         MOVES CON
15.         GRASP 25
16.         DEPART 50
17.         MOVE PALLET:APP
18.         MOVES PALLET
19.         OPENI
20.         DEPART 50
21.         SIGNAL GOCON
22.         SETI PX = PX + 1
23.         RETURN
24.   40    STOP
```

Figure 8–14   Palletizing With VAL.

■ Work-cell simulation software or a robust robot programming language must be available to program the cell off-line.

The first condition rarely occurs in standard industrial robots. For example, when a robot is programmed off-line to go to a translation point in the work envelope, it misses the point because of the mechanical tolerances in the arm linkages and feedback mechanisms. Effective off-line programming is only possible with highly accurate placement of work-cell hardware and calibration of the robot arm. This type of robot and work cell calibration is possible with integrated systems such as WORKSPACE and ROBOTRAK from Robotic Workplace Technologies Inc. ROBO-TRAK provides the calibration of the robot arm and work-cell hardware that is used in the WORKSPACE program. WORKSPACE is simulation software that builds a manufacturing automation simulation and integrates the work-cell equipment location data and robot arm signature captured with ROBOTRAK. The robot signature data include the zero position of each joint, the length of each link, the distance offset at each link, and compliance at each joint. These types of systems measure static position and motion paths to an accuracy of 0.2 millimeters (0.008 inches) in three dimensions. As a result, a functional off-line program is developed with WORKSPACE by integrating the location data for all work-cell devices and compensating for the variations present in the mechanical linkages in the robot.

Translation Points and System Variables

The translation points, step 5 in the programming steps, are created using one of the five methods described earlier. Method 1 is an *on-line* programming technique in which visual judgment establishes the fit and alignment between the tooling and parts. Although it requires a stop in production, it is the most frequently used method for establishing the translation points in a cell. Methods 2 and 5 are *off-line* programming techniques for establishing translation points. In each case, the translation point is established using coordinate data relative to the robot reference frame. Therefore, these techniques require some system to accurately measure the position of work-cell hardware and calibrate the robot arm for positional accuracy.

The creation of the translation points is the most costly part of the programming if production must be stopped; therefore, it is important to teach as few points as possible for a given program. Using *base* or *reference* points the remaining position points in the program can be calculated using the process described in method 3.

The motion variables associated with translation points in Figure 8–13 include *velocity, tool center dimensions, Cartesian coordinate values, language functions,* and *commands.* These variables are usually included on the task point graph as a part of the program development.

CASE: CIM AUTOMATION AT WEST-ELECTRIC

Components for the automated upset forging cell are in fabrication. The bowl feeder and other material-handling equipment were specified and are ready to go to the vendor. The gaging system was bid and came in under the budgeted amount. A spare oven was modified to test the positioning accuracy of the parts

1. Load oven with parts at 15 second intervals
2. Start upset forging operation
   a) Get part from parts feeder (Conditions: parts in feeder, gripper open)
   b) Put cold part in oven (Conditions: part in gripper, oven door open, table in same position after last part was removed)
   c) Remove heated part from oven (Conditions: gripper open, table indexed one position,)
   d) Place heated part in forge press (Conditions: oven door closed, part in gripper, press in open position, previously forged part ejected from die)
   e) Cycle forge press (Conditions: part seated in die hole, gripper clear of press)
   f) Repeat manufacturing cycle
3. Modify manufacturing cycle at 100th part to measure part parameters.
4. Continuously check oven temperatures, parts feeder levels, safety screen, emergency stop buttons, and condition of switches on the operator interface panel.

Figure 8–15 *Cell Operational Sequence.*

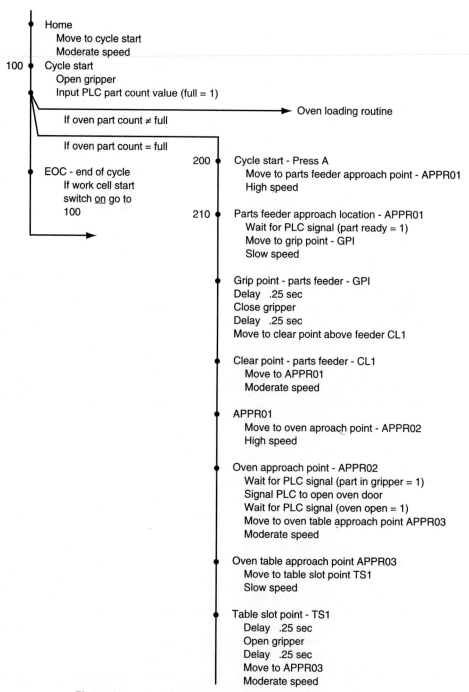

Home
    Move to cycle start
    Moderate speed

100   Cycle start
    Open gripper
    Input PLC part count value (full = 1)

                                           → Oven loading routine

    If oven part count ≠ full

    If oven part count = full

                      200   Cycle start - Press A
EOC - end of cycle   Move to parts feeder approach point - APPR01
    If work cell start   High speed
    switch <u>on</u> go to
    100   210   Parts feeder approach location - APPR01
                                    Wait for PLC signal (part ready = 1)
                                    Move to grip point - GPI
                                    Slow speed

                                  Grip point - parts feeder - GPI
                                  Delay   .25 sec
                                  Close gripper
                                  Delay   .25 sec
                                  Move to clear point above feeder CL1

                                  Clear point - parts feeder - CL1
                                  Move to APPR01
                                  Moderate speed

                                  APPR01
                                  Move to oven aproach point - APPR02
                                  High speed

                                  Oven approach point - APPR02
                                  Wait for PLC signal (part in gripper = 1)
                                  Signal PLC to open oven door
                                  Wait for PLC signal (oven open = 1)
                                  Move to oven table approach point APPR03
                                  Moderate speed

                                  Oven table approach point APPR03
                                  Move to table slot point TS1
                                  Slow speed

                                  Table slot point - TS1
                                  Delay   .25 sec
                                  Open gripper
                                  Delay   .25 sec
                                  Move to APPR03
                                  Moderate speed

Figure 8–16   Initial Points in the Upset Forging Task Point Graph.

table, and the testing is planned in a week. Bill is satisfied that they will have a successful work cell. One of the last tasks is the development of the robot and PLC cell control program. He asked Jerry to develop an operational sequence (Figure 8–15) for the work cell in preparation for developing a PLC ladder logic diagram. Bill worked with Mike on the task point graph for the robot (Figure 8–16). After feedback from the team on each of these items, Ted and Mike would add the data to the WORKSPACE simulation software to test the process. ●

8-10 SUMMARY

The computers used in robot systems are responsible for four functions: machine manipulation, sensing, logical decision making, and data processing. The programming languages currently used to execute these four functions have no common format or standard form. At a higher level, work-cell control software manages the flow of information between cell devices and the enterprise. Work-cell software falls into one of three categories: in-house developed systems, application enablers, and adoption of an open system interconnect. One level below the cell control software, programmable logic controllers or robot controllers control sequential cell activity.

There are as many robot programming languages as there are robot manufacturers. The robot languages were developed by means of three techniques. The first method starts with the design of a pure manipulator control language and adds the other high-level language features as demand requires. The second adopts an existing high-level language and augments it with the manipulator commands required for robot operation. The third technique includes the design of a high-level language for robot control and support of other external data processing functions.

The many current languages can be classified into four groups based on the level at which the user must interact with the system during the programming process. The four groups are joint control languages, primitive motion languages, structured programming languages, and task-oriented languages. At present, the most frequently used languages are in the structured programming category.

Reference frames are sets of Cartesian coordinates that describe the positional relationship between the robot, the gripper, the tool center point, the workpiece, and the universe in which all of these exist. The dimensional relationships between the parts of the robot and the production parts in the work cell are described by means of mathematical equations called transformations. Linking the robot motion to the reference frames using mathematical transformations permits sophisticated control of robot motion and off-line programming.

In on-line programming, the production system must stop to allow the robot programmer access to the production robot for program generation or modification. The lost production time is reduced if some of the program development is performed off-line and downloaded to robot before teaching the position points.

True off-line programming is performed away from the production area on work-cell simulation software. Off-line programming requires a work-cell hardware and robot calibration process.

Five basic methods are available to define translation points for servo robots: (1) teaching of translations by robot positioning; (2) teaching of translations by coordinate input; (3) creation of new translations by algebraic expressions; (4) creation of a new translation by use of an external interrupt; and (5) downloading new translations through the RS-232C port. Translation points for nonservo robots are limited to the fixed stops provided by the system. Nine steps are used to develop a robot program: (1) establish a basic program structure; (2) analyze the manufacturing process in which the robot will work; (3) divide the robot action into tasks and subtasks; (4) draw a task point graph that describes the desired motion; (5) identify the translation points on the task point graph; (6) assign values for all system variables to control the motion; (7) write and enter the command statements; (8) create or teach the translation points; and (9) test and debug the program.

QUESTIONS

1. Describe the four system functions performed by the computer in high-technology robots.
2. Describe the three categories of work-cell controller software.
3. Describe the differences in programming robots controlled by PLCs and those with special-purpose controllers and languages.
4. What are the four levels of robot languages?
5. What are the advantages and disadvantages of languages at each level?
6. How do Grafcet and ladder logic programming for PLCs differ?
7. What is the function of the CYCLE START and END OF CYCLE programmed points in a robot program? Where is the optimum position for the CYCLE START point?
8. What is a TPG and what information is included in the graph?
9. Describe the two parts of a robot program for servo and nonservo robots.
10. Describe reference frames and explain how they are used in the DARL language for the Seiko robot.
11. Describe the difference between on-line and off-line programming.
12. Describe the five methods used to program a translation point for a robot program. What program development situation or work-cell condition would dictate the use of each method? Which method is used most often on robots in operation now?
13. How do the *WORLD, TOOL, JOINT,* and *FREE* operational modes differ? Identify the mode(s) that would be used for each of the following situations:
    (a) Teaching move points between fixtures in the work cell
    (b) Rotation of the tooling from one side of the work cell to the other with no change in the elevation or radial extension of the arm

(c) Movement along the row of a pallet when the pallet coordinates are aligned with the robot coordinates

(d) Movement to insert a part in a fixture plate when the fixture coordinates are not aligned with the robot coordinates

(e) Separation of the robot tooling from the fixture after a system fault caused the controller to shut down in the middle of an assembly operation

14. Describe the steps used in developing a robot program.

15. How are program tasks and subtasks identified? How is each handled in the robot program?

PROBLEMS

1. Modify the task point graph in Figure 8–13 to unload two production machines instead of one.

2. Develop a ladder logic diagram for each of the nonservo robot motions (A, B, C, D, and E) in Figure 8–9. Assume the system is interfaced to a PLC and assign necessary input and output module identification. Put in a 2 second delay at each position.

CASE PROJECTS

1. Develop a task point graph for the extrusion work cell developed in question 2 of the case/projects section in Chapter 4.

2. Develop a chart showing the operational sequence for the extrusion work cell developed in question 2 of the case/projects section in Chapter 4.

3. Draw a ladder logic diagram for the operational sequence chart developed in question 2. Use input and output module notation and functions for any available PLC.

4. Draw a ladder logic diagram for the W-E upset forging operational sequence. Use input and output module notation and functions for any available PLC.

Justification and Applications of Work Cells

9-1 INTRODUCTION

Manufacturing is a collection of interrelated activities that include product design and documentation, material selection, planning, production, quality assurance, management, and the marketing of goods. The fundamental goal of manufacturing is to use these activities to convert raw materials into finished goods efficiently, in a timely manner, and on a profitable basis. Every action taken in manufacturing must be justified based on its effect on this fundamental goal. Peter Drucker, a noted educator, writer, and management expert, wrote in *The New Realities:*

> We have known for a long time that there is no one right way to analyze a proposed capital investment. To understand it we need at least six pieces of information:
>
> ■ Expected rate of return
> ■ Payout and the investment's expected life
> ■ Discounted present value of all returns through the productive life of the investment
> ■ Risk in not making the investment or deferring it
> ■ Cost and risk in case of failure
> ■ Opportunity cost that is the return from alternative investments
>
> Every accounting student is taught these concepts. But before the advent of data-processing capacity, the actual analyses would have taken man-years of

clerical toil to complete. Now anyone with a spreadsheet should be able to do them in a few hours. The availability of information thus transforms the capital investment analysis from opinion to diagnosis, that is, into the rational weighing of alternative assumptions. Information transforms the capital-investment decision from an opportunistic, financial decision, governed by the numbers, into a business decision based on the probability of alternative strategic assumptions. As a result, the decision both presupposes a business strategy and challenges that strategy and its assumptions. What was once a budget exercise becomes an analysis of policy.

Manufacturers face enormous internal and external challenges because of the increased use of technology in the production and marketing of goods and services. As a result, manufacturing strategies such as order-winning and order-qualifying criteria, described in Chapter 1, are needed to link manufacturing and management. A focus on improving the production standards, such as setup time, inventory, quality, machine uptime, and manufacturing space ratio, makes manufacturing the strategic difference in the quest for greater market share. As Drucker noted, a business strategy is assumed, and the way capital invested as a result of this strategy is justified has changed significantly. Justification must now go beyond the tangible numbers.

9-2 CAPITAL EQUIPMENT JUSTIFICATION

The principal considerations for the justification for capital expenditures in manufacturing can be expressed in the following equations:

$$\text{payback period} = \frac{\text{total investment}}{\text{total yearly savings}}$$

$$= \frac{\text{total investment}}{\text{savings} - (\text{saving} \times \text{TR}) + (\text{DP} \times \text{TR})}$$

The total investment includes all costs associated with the automation, savings include yearly savings which result from the automation, TR is the tax rate, and DP is the yearly depreciation for the capital equipment. The following examples illustrate the *payback method:*

- *Given:* Total investment is $250,000, savings per year is $105,000, tax rate is 33 percent, and the depreciation on equipment is $28,000.
- *Find:* Payback period

$$\text{payback period} = \frac{\$250,000}{\$105,000 - (\$105,000 \times 0.33) + (\$28,000 \times 0.33)}$$

$$= \frac{\$250,000}{\$79,590}$$

$$= 3.14 \text{ years}$$

Return on Investment Method

Return on investment (ROI) is the percentage return on the automation investment. The ROI is the ratio of savings to investment, in other words, the reciprocal of the payback period:

$$ROI = \frac{1}{payback\ period}$$

The terms in the equation are the same as those defined in the payback calculations. ROI is calculated as follows:

■ *Given:* The payback period is 3.14 years.
■ *Find:* ROI

$$ROI = \frac{1}{3.14}$$

$$= .318\ or\ 31.8\ percent$$

Cash Flow Method

Cash flow analysis is a variation of the payback technique. Stated as an equation,

$$cash\ flow = \Sigma(positive\ and\ negative\ cash\ values\ over\ time)$$

As noted in the statement of Drucker, spreadsheet software on microcomputers gives the designer of the work cell the power to make investment decisions and do what-if analysis on the alternatives for automation investments. The cash flow method is an ideal application for spreadsheets. Study the cash flow spreadsheet in the West-Electric case later in the chapter.

Time Value of Money

The time value of money describes the change in the value of an investment or a savings in the future due to monetary inflation and interest. For example, as a result of inflation, $1000 invested today has a greater value than $1000 invested 2 years hence. Similarly, a savings of $1000 expected 3 years after automation is implemented is worth less than $1000 in present dollars because of the interest the money could earn in 3 years. The benefits of automation must always be justified by future events that are affected by the time value of money. Therefore, the economic analysis methods discussed earlier must be discounted for the future events. Often the equations are rewritten as follows:

$$payback\ period = \frac{net\ present\ value\ of\ \Sigma\ (investments\ over\ time)}{net\ present\ value\ of\ \Sigma\ (savings\ over\ time)}$$

$$discounted\ cash\ flow = net\ present\ value\ of\ \Sigma\ (positive\ and\ negative\ cash\ values\ over\ time)$$

The term *net present value* means that all future dollar investments and savings are discounted, or converted, to their *present worth*. This process requires an estimate of inflation (average or yearly) and market interest rates (average or yearly) for future periods in the justification. Return on investment equations that include discounted cash flow analysis are provided in many financial analysis texts and will not be discussed in detail here. Consult the justification spreadsheet in the West-Electric case; it provides a discounted cash flow analysis that includes the impact of the discount or interest rate on money but does not include an inflation factor.

Justifying Robotics Applications

The justification principles described in this chapter for capital expenditures in manufacturing apply equally well to robot-driven automation. The cost of the automation investment must be justified with future savings and revenue. There are, however, several distinct differences when robots are involved. The introduction of robots implies the displacement of human operators; thus, a significant saving occurs in direct labor cost. Robots usually increase productivity and output because of uniform production rates and lower production cycle time. Robots enhance product quality and uniformity and support flexible and agile manufacturing. Finally, the reprogrammability of the robot gives the investment in robotics a longer lifetime, because the robot is reusable when production on the current product is stopped. These factors take on even greater significance when a breakdown of the investment dollars in robotic work cells is analyzed. The robotics application cost graph in Figure 9–1 shows the percentage of investment dollars in accessories, basic robot system, and installation costs. In each case the basic robot cost is the largest single investment. Therefore, the largest part of the automation investment is in a machine that has a production lifetime longer than other hardware in the work cell.

9-3 AUTOMATION APPLICATIONS

Automation applications in industry vary widely; however, most robot applications would fall into one of the six categories listed in Figure 9–1. In this section, each of the applications areas are described.

Material Handling

Material handling generally involves the movement of material through manufacturing. The automation technologies used for this work include conveyors, automatic guided vehicles (AGVs), and servo and nonservo robots. The product size, production layout and environment, and flexibility needed in the application determine if a robot is used and what type is selected. The automotive, foundry, light and heavy manufacturing, and electronics industries use large numbers of robots for material handling.

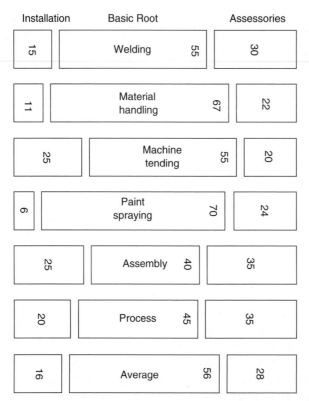

Installation	Basic Root	Assessories
15	Welding 55	30
11	Material handling 67	22
25	Machine tending 55	20
6	Paint spraying 70	24
25	Assembly 40	35
20	Process 45	35
16	Average 56	28

Figure 9–1 Percentage Cost Breakdowns for Robotic Cells.

APPLICATION NOTE
MATERIAL HANDLING OF ELECTRICAL RELAYS

Feme improved product quality and achieved a labor savings of two and one-half operators per shift by automating its electrical relay calibration and testing process. The system has a payback period of 1 year.

The two work cells (Figure 9–2), designed by Feme's Industrial Automation Division, consist of an AdeptOne robot and three identical calibration and test stations. The AdeptOne feeds relays in batches of 500 or more into the test stations. There are 120 different relay models differing by coil voltage, contact configuration, and number of poles. The cells run two or three shifts per day, 5 or 6 days per week, depending on market demand.

System Configuration

In each cell, the AdeptOne robot stands in the center surrounded by three test and calibration stations, and a conveyor. The conveyor delivers pallets of untested relays to the robot and removes the pallets after testing and calibration are complete. Pallets for holding rejected relays are on a pneumatic slide next to the conveyor. The robot is equipped with a two-fingered gripper.

Work Cell Operation

An operator loads the pallets containing thirty-six relays onto the conveyor. When the pallet arrives to the pickup position, the robot unloads a single relay and transfers it to a test station. The robot repeats the operation twice more to fill all three stations. The relays are picked up in a vertical position. During the transfer the relays are turned 90 degrees to a horizontal position for feeding into the test stations.

The length of the test and calibration cycle varies between 5 and 7 seconds, depending on the performance of each relay. At the end of the cycle, the test station communicates the results to the robot. If the relay is acceptable, the robot removes the relay from the test station and returns it to the same pallet on which it entered the cell. If it is rejected, the test station also communicates which of six classes of failure the relay experiences and the robot then removes the relay from the test station and places it on the "reject" pallet classified by the type of failure.

Changeover between relay models simply requires the operator to select a different program at the terminal. The gripper is designed to accommodate all 120 models.

By automating its relay test and calibration process, Feme has improved its product quality and reduced its labor costs.

(Courtesy of Adept Technology, Inc.)

Figure 9–2   Electrical Relay Calibration Cell. (Courtesy of Adept Technology, Inc.)

APPLICATION NOTE
COOKIE PACKAGING

Kambly, a manufacturer of cookies located in Trubschachen, Switzerland, sought an effective solution to the rising cost of cookie assembly. Dedicated automation was ruled out since a high level of flexibility and precision was required to assemble a wide range of delicate products (Figure 9–3).

A robotic solution proposed by Adept System Integrator A.R.T. (formerly Adec Robot AG), provided Kambly with the answers they sought. The flexible automation system installed by A.R.T. provided Kambly with the desired results—improving quality, lowering costs, and increasing production line flexibility.

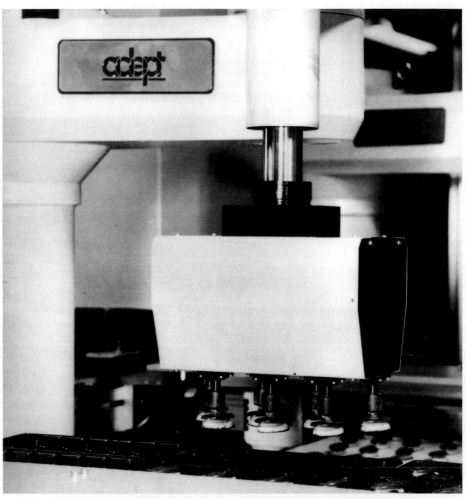

Figure 9–3   AdeptOne Robots With AdeptVision AGS-GV Package Cookies. (Courtesy of Adept Technology Inc.)

System Configuration

Kambly's automated packaging system is composed of four AdeptOne robots working with AdeptVision AGS-GV. The robots line each side of a conveyor that transports cookies from an oven. Each robot is equipped with a vacuum gripper that can delicately handle either seven or nine cookies, depending on the batch in production.

The robot work cell features an integrated vision system with four cameras viewing the conveyor from below. The cameras are not only responsible for guiding the robots, but also perform a variety of quality control functions.

Work Cell Operation

Cookies arrive at the robotic packaging cell randomly distributed along a 1-meter wide conveyor belt, traveling at 5 meters per second. The first pair of cameras locate the cookie's position and guide two upstream robots to track the cookies on the line. A seven- or nine-cup suction gripper acquires the cookies from the moving line, then packages the cookies in a blister pack on an adjacent pallet conveyor. The second pair of AdeptOne robots, also guided by two cameras, repeat the sequence to complete the package. Two cycles are necessary to fill the fourteen- or eighteen-cookie blister packs.

But the vision system does more than just instruct the robot how to pick up and orient a cookie. It also inspects the size of the product. If the cookie is not within the required limits, the cookie is left to fall off the end of the conveyor into a reject bin. In normal operation, Kambly packages four different types of cookies while operating the Adept robot system two shifts a day, 5 days a week.

Changeover of the production line takes between 15 and 30 minutes, providing Kambly the flexibility to produce a wide variety of products. A.R.T. designed the end of arm tooling so that changeover is accomplished by replacing just four screws.

The requirements of Kambly are not unique to their industry as most companies are seeking to maximize production flexibility, improve quality, and lower production costs. If an integrated system featuring robots, vision, and flexible feeding enabled Kambly to achieve these production results, perhaps the Adept approach should be part of your "recipe" for success.

(Courtesy of Adept Technology, Inc.)

Machine Tending

In most machine tending applications the robot replaces the human operator. The operation usually includes the loading of raw material into a production machine and the removal of the finished part when the production process is complete. In other cases, such as die casting and plastic injection molding, the robot just unloads finished parts from the machine after the molding or casting process is complete. In most applications, servo robots are used because adequate payload, large work envelope, and dexterity are not available in pneumatic stop-to-stop machines. The robot in Figure 4–19, which loads and unloads a turning center, is a good example of machine tending. Note that the robot is loading the machine from the rear, so that manual operation from the front is possible if the robot system is not operational. The West-Electric case study provides another example of machine tending with robot technology.

Assembly

Assembly applications occur in all five manufacturing systems: project, job shop, repetitive, line, and continuous. The level of complexity varies from highly intricate (the assembly of high-tolerance mechanical parts) to simple (the construction of walls in residential construction). In addition, the volume varies from low to very high. As a result, the window of opportunity for robotic assembly falls in the middle. Servo robots are primarily used in assembly tasks for medium-volume repetitive and line-type production (Figure 2–17). Stop-to-stop robots are integrated into fixed automation machines (Figure 1–7) and systems and are used in high-volume line and continuous production areas. Robotic assembly is heavily integrated into the production of electronic parts and products, but only used lightly in the automative and the light manufacturing sectors.

APPLICATION NOTE
MECHANICAL ASSEMBLY OF FLUORESCENT LIGHT FIXTURES

Zumtobel, located in Dornbirn, Austria, has integrated an AdeptOne robot cell into its automated line, producing fluorescent lighting units at a rate of one every 30 seconds (Figure 9–4).

A robotic solution was chosen because of its unique ability to manage a high production rate and mix of products. Adaptable to forty variations of product, the Adept robot is

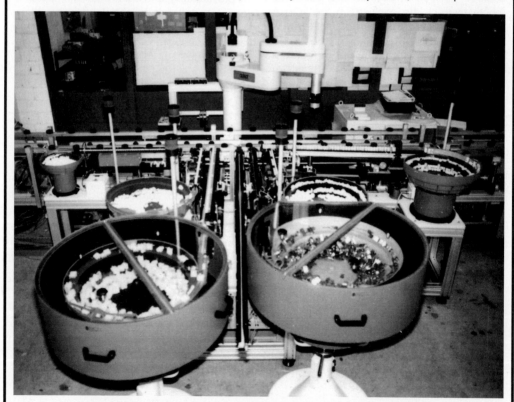

Figure 9–4 Automated Assembly of Fluorescent light Fixtures. (Courtesy of Adept Technology, Inc.)

responsible for assembling an average of twelve components into each lighting unit.

System Configuration

The AdeptOne robot cell, designed by A.R.T., is completely integrated with Zumtobel's automated line for manufacturing fluorescent lighting units.

A continuous conveyor passes through the center of the work cell delivering fluorescent light housings to the robot for assembly. Once the mechanical components have been assembled, the lighting units are transported to dedicated stations that add the electrical elements.

Six bowl feeders located opposite of the robot, orient and position the six different components that the robot has to assemble.

Altogether, the AdeptOne robot is programmed to assemble forty different variations of fluorescent lighting tubes. The Adept MC Controller, linked to a VAX computer, is responsible for downloading to the Adept cell which product to assemble. At the same time, the MC Controller sends progress reports back to the VAX computer.

Work Cell Operation

As each unit arrives at the robot assembly station, it is grasped and centered by an indexing pin.

The movements of the indexing pin are controlled by the MC Controller, acting as an external linear axis. In order to minimize the length of each robot movement, the pin moves the lighting unit back and forth during the assembly cycle.

As assembly needs change with future lighting units, various grippers will be required. Therefore, system integrator A.R.T. provided the cell with a gripper change facility.

By installing flexible automation, Zumtobel has been able to completely automate the assembly of its lighting units and reduce required manpower. Altogether, forty different product designs can be assembled at a rate of one every 30 seconds.

(Courtesy Adept Technology, Inc.)

APPLICATION NOTE
MECHANICAL ASSEMBLY OF DISHWASHER AND VACUUM CLEANER COMPONENTS

The Miele factory in Bielefeld, Germany, installed a flexible automation solution using a single AdeptThree robot (Figure 9–5) to assemble two different products: the steam exhaust unit in a dishwasher and the device in a vacuum cleaner that signals when the dust-collecting bag is full. Installation of the automated system has resulted in a labor savings of two and one-half operators.

The system, designed by Proline GmbH, is capable of running almost completely unattended, three shifts per day, and has a production rate of one finished product per 30 seconds.

System Configuration

The AdeptThree robot stands in the center of the cell, surrounded by part feeding equipment and assembly workstations. The dishwasher steam exhaust assembly consists of seven parts. Specially designed molded plastic trays, containing four of each of the seven parts, are stacked in a vertical magazine. Each tray is automatically removed from the magazine and transferred by a conveyor system to the robot pickup position. Completed assemblies exit the workstation by a conveyor system parallel to the first. Two assembly fixtures are located on either side of the robot's base.

The AdeptThree robot is equipped with a triangular end-effector with suction cups at two corners, an expanding three-finger gripper on the third corner, and in the center, a parallel gripper.

Work Cell Operation

The assembly process begins when the robot picks up a wheel, a mounting plate, and a motor from the plastic tray, and assembles them on one of the two fixtures.

Next, the robot picks up three more components: a rubber seal, a small wheel, and a housing. It assembles the seal and the housing on the other fixture. It then moves over to a screwdriving station just behind its base and loads first the small wheel followed by the housing assembly, and the motor, to the screwdriving fixture. The robot then loads a temperature regulating unit onto the assembly and the screwdriving operation takes place. The Adept robot removes the finished product and places it on a tray. Once the tray is full it is transferred by conveyor to a vertical magazine.

To assemble the vacuum cleaner product, the robot changes its gripper, and follows a similar stacked-assembly procedure.

In specifying the design of the cell, Miele has attempted to achieve the greatest degree of autonomy possible. If there is a problem with one of the assembly tasks, the robot will repeat the operation three times before rejecting the problematic part. It will then take a new part from a reserve located just behind the robot base and try the operation three additional times. If it is still unsuccessful, the robot is programmed to remove all the components of a started subassembly from the different fixtures and continue with a fresh set of parts.

(Courtesy Adept Technology, Inc.)

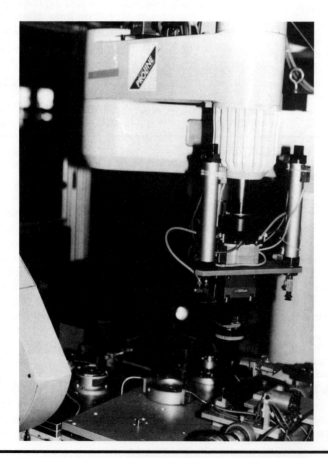

Figure 9–5 Assembly of Dishwasher and Vacuum Cleaner Components. (Courtesy of Adept Technology, Inc.)

An automotive manufacturer using robots to insert pistons into engine blocks wanted to check for conditions such as missing piston rings, jammed rings, or collisions between pistons and crankshafts.

To perform this task, an Assurance Technologies robotic F/T sensor was mounted to a Seiko RT-5000 robot (see Figure 9–6). A three-finger gripper, mounted on the end of the F/T sensor, was designed to close around the piston rings, centering and compressing them slightly while grasping the body of the piston.

The piston has an oil ring, a bottom compression ring, and a top compression ring. As each is pushed into the cylinder, the insertion force rises until the ring is in place; it then decreases to the level required to push the piston itself into the cylinder. This change in force is detected by the F/T sensor and analyzed by an IBM AT computer to determine the presence or absence of each ring. The F/T controller provides data to the computer via an RS 232 serial port. The controller's discrete I/O port is connected to the robot to provide force threshold communication, and passes this information through the I/O port to the robot. Thresholding detects jamming, collisions between connecting rod and crankshaft, dropped pistons, and crashing.

As part of the analysis, a time derivative of the force signal is calculated and used to detect the rings as they snap into the cylinder during insertion. Using the time derivative rather than the signature itself simplifies the ring detection problem by eliminating the commonly encountered variations in absolute insertion force. Such variations are caused by slightly different insertion conditions for each piston/cylinder combination.

The piston rings appear as zero-crossing spikes in the derivative signal. The compression rings generate single spikes. The oil ring, consisting of a plastic spacer sandwiched between the steel rings, generates a double spike. The derivative signal is compared to a predetermined threshold value to detect presence/absence of a ring.

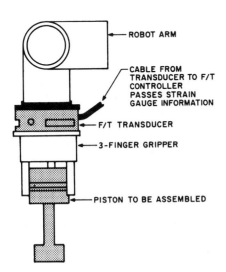

ROBOT ARM

CABLE FROM TRANSDUCER TO F/T CONTROLLER PASSES STRAIN GAUGE INFORMATION

F/T TRANSDUCER

3-FINGER GRIPPER

PISTON TO BE ASSEMBLED

Figure 9–6 Force/Torque Sensing in Piston Assembly. (Courtesy of *SENSORS—The Journal of Machine Perception*)

Figure 9–7 shows the data obtained during the insertion of a piston with the bottom compression ring missing. This condition is detected by the analysis software, which produces an error condition that prevents the piston assembly from being passed along the line. The robotic force sensor, which in effect provides 100 percent inspection, dramatically improves the quality of the piston insertion procedure.

(Reprint with permission from SENSORS—*The Journal of Machine Perception,* November, 1992.)

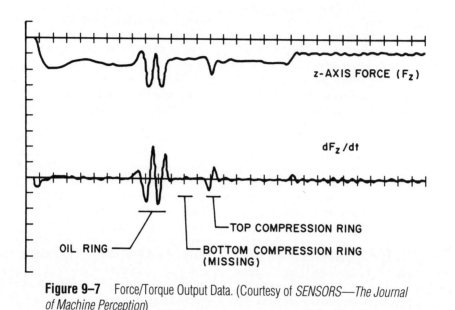

Figure 9–7 Force/Torque Output Data. (Courtesy of *SENSORS—The Journal of Machine Perception*)

Process

All applications that use the robot as a production machine are grouped into this classification. The processes performed frequently by a robot include welding, painting, sealing, deburring, grinding, drilling, and finishing. Welding and painting are major application areas for the robotic industry. Deburring of raw casting, a process application developed by ABB, is illustrated in Figure 9–8. Another process area, the application of glues and sealers, is time-consuming and tedious when performed by hand; as a result, industrial robots easily outperform their human counterparts. A robot offers the advantages of accurate bead placement and a high application rate. The consistency in rate of application provides an additional savings in material. Figure 9–9 shows a robot with a three-roll wrist applying sealer to a sheet metal joint in the passenger compartment of an automobile. Note the ability of the robot to enter the car through the window of the side door. In some cases, the robot can apply the sealer while the car continues to move on the conveyor.

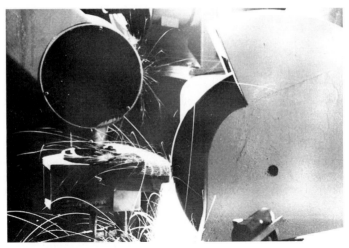

Figure 9–8   Robot-Grinding Application. (Courtesy of ASEA, Inc.)

Figure 9–9   Sealer Application.

At Pepperidge Farm's award-winning plant in Denver, Pennsylvania, high-risk, repetitive motion has been reduced in the cookie sandwiching operation. Milano and Brussels cookies are now sandwiched by twenty AdeptOne robots (Figure 9–10) with AdeptVision resulting in an annual labor savings of close to $1 million through a reduction of eight jobs per shift.

Several years ago, Pepperidge Farm determined that cookie sandwiching posed a high risk of carpal tunnel syndrome. Pepperidge reviewed various automation options and eventually selected robotics, because it was the only solution capable of handling the wide range of products required.

The time required from the start of baking to finished packaging is 40 minutes.

System Configuration

The system, designed by system integrator Technistar of Longmont, Colorado, consists of four AdeptOne robots with HyperDrive and sixteen standard AdeptOnes, each equipped with AdeptVision AGS-GV. Ten robots stand on each side of a moving conveyor.

Work Cell Operation

Cookies are mixed, deposited, and baked twenty across on a 1-meter wide moving conveyor. After exiting the oven, the outside lanes of cookies travel over a chocolate bottomer

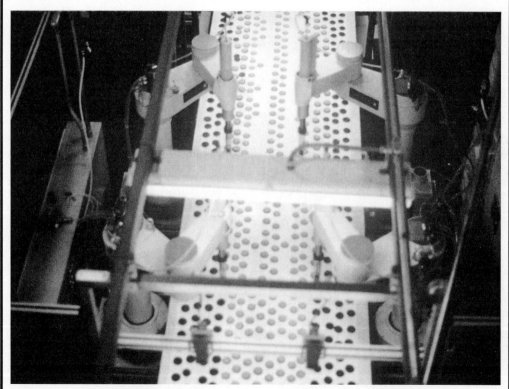

Figure 9–10   Cookie Assembly. (Courtesy of Adept Technology, Inc.)

while the center ten lanes travel over a bypass conveyor. The chocolate-coated cookies turn upside down by falling out of the chocolate bottomer onto the main conveyor. Cookies from the bypass conveyor rejoin the chocolated cookies on the main conveyor before entering the robot work cell.

The robot picks up the non-coated half from the center of the conveyor and places it on the chocolate-covered half. AdeptVision AGS-GV finds the randomly oriented cookies and guides the robot to pick up an uncoated half and wait for an unsandwiched coated half to come by on the conveyor. The robots at the beginning of the line work faster than those at the end because more unsandwiched cookies are available. After the sandwiching station, the cookies move into a cooling tunnel then on to handpacking in paper cups. The cups of cookies are then automatically bagged, weighed, and then manually packed into shipping boxes.

In the future, Pepperidge Farm plans to continue automating high-risk operations. In the near term, it plans to evaluate automating its cookies cupping operation where cookies are manually placed in paper cups in preparation for bagging.

(Courtesy Adept Technology, Inc)

Welding

Welding applications are classified in three categories: resistance, arc, and gas. In each application area only servo-type robots are used because path control is critical. Resistance or spot welding was developed by the automotive industry in the mid 1970s and continues to be a major application in that industry. The robot became a standard item on the assembly line due to its ability to effortlessly move the 150-pound welding gun into position on a moving car body and produce accurate resistance welds at a rate of one every 3 seconds. Figure 9–11 shows a robot performing spot welding on a moving car body. Robotic arc welding is frequently used in the heavy-equipment manufacturing industries to produce off-road hauling and earth-moving equipment. Lighter applications occur in the automotive, electronics, and aerospace industries.

Metal inert gas (MIG) and tungsten inert gas (TIG) welding with real-time interactive control permit the robot to track the seam between two metal plates. The seam to be welded is located and tracked by a photo sensor, such as the vision camera in Figure 9–12, or by welding parameters. The location of the seam or welding path is passed to the robot controller, which moves the arm holding the welding tooling. The location and control are dynamic, so that corrections in the direction of the welding gun are made as the weld is in progress. Robotic arc welding is cost-effective when compared with manual and semiautomatic systems. For example, a comparison of cycle times indicates that manual and semiautomatic welding are 30 and 45 percent efficient, respectively, whereas robotic arc welding is close to 70 percent efficient. In robot welding systems the robot controller has the capability to control all the welding variables including *arc voltage, wire feed rate,* and *torch speed.* In addition, the robot system can control the position of a multi-axes part positioner, like the one visible in Figure 9–13.

Gas welding is used to accurately cut complex curves and shapes in pipes and flat metal plates. Although it is not used as frequently as spot and arc welding, it is a major robot application.

Figure 9–11 Spot Welding Operation on Moving Car Body. (*Source:* Rehg, James A., *Computer Integrated Manufacturing.* Englewood Cliffs, NJ: Prentice Hall; 1994, p. 350)

Painting

In general, paint-spraying applications require limited automation technology. The number of sensors required is relatively low, the support equipment is primarily material handling, and robot precision is not critical. Servo paint-spraying robots have traditionally cost more than the standard servo robot, but the cost of the work cell is about the same because the support hardware is less. Several applications are supported by robotics:

- *Dip coating:* The part is lowered into the coating material reservoir and removed to allow excess material to flow off the part. In some cases parts are spun to remove excess coating material. A variety of robots are used in this application area, including servo-and nonservo-type arms.

Figure 9–12 MIG Weld Tooling (Right) and Weld Vision System (Left) on Common Tool Plate Attached to Three-Roll-Wrist Robot.

Figure 9–13 Welding Application With Part Positioner.

Peacocks and robots are not exactly birds of a feather, but at Kendon Inc., ABB Robotics' IRB 1500 is helping to turn out beautiful peacocks for the home interior products market.

Kendon Inc., a 21-year-old company based in Winnsboro, Texas, manufactures metal sculpture wall decor items and markets them through Home Interiors Inc. of Dallas and through Kendon's wholesale distribution division, "American Accents by Ditto." Its popular peacock is a 4-foot tall, brass-plated sculpture, made of 3/16-inch wire, which boasts a large plume that posed a challenge for the IRB 1500 robot. In fact, according to Shane Wilson, Kendon's new products coordinator, they bought the robot with the peacock—and its unusual tail—in mind.

Last year when the new product was to be put on line, Wilson and CNC Systems Coordinator Dan Cartwright started looking at robot systems because they knew automated welding would be necessary to make production cost-effective. After looking at a number of different robots and visiting the 1993 AWS show in Houston, they selected ABB Robotics' IRB 1500.

The IRB 1500 is designed to ensure a minimum of maintenance and service. Its flexibility and speed make it ideal for precision arc welding, and its S3 control system can handle six internal robot axes and six external axes. Wilson notes that the features which "cut the cake" for them were the robot's portable teach pendant and the joystick used for programming. The joystick is not only easier to use than push buttons; it is also 25 percent faster.

Once they made their decision, Wilson and Cartwright moved quickly. They purchased the IRB 1500 straight from the AWS Show floor in Houston, had it installed the following week, had their people working on building a positioning table and part-holding fixture, and headed for ABB Robotics' Robot University training course at the Welding Systems Division headquarters in Fort Collins, Colorado. When they returned, everything was ready to go. "I'd say we were in production within about 5 weeks," recalls Wilson.

The plume, or tail section, welded by the robot (Figure 9–14) is about a foot wide by 2 feet long. It consists of thirteen pieces, each about 6 inches long, that look like fish hooks. These pieces are all placed on a fixture, and the robot makes the peacocks' joints by connecting them with about fifty small tack welds. The positioning table has two stations, so that as the robot welds on one side, the operator can load and unload parts on the other side. When welding is complete, the table indexes and presents the next workpiece to the robot. According to Cartwright, the robot has reduced the welding time of approximately 6 minutes to 1 minute, 48 seconds.

Once the plume is completed, it is manually welded to the head and breast subassemblies, and the entire peacock is then brass plated. The entire process takes about 11 minutes from start to finish.

The robot operates for one 10-hour shift, 4 days per week and the plant, which employs about 200 people, currently produces 1100 to 1500 peacocks per week. Cartwright handles most of the programming, and he and Wilson have trained three other employees to operate the robot. Those employees are also able to handle occasional editing of the welding program.

The company plans to phase out the peacock sculpture, so it is currently in the process of designing a new product that will take advantage of the robot's capabilities. While there are no plans for another robot at this time, according to Wilson: "If we get a new product on line for the robot and it does well, you never know. If we had the need, we'd definitely get one, and it would more than likely be another ABB robot."

Cartwright agrees, adding: "Our purchase of the ABB robotic welder is indicative of Kendon's philosophy; we believe in manufac-

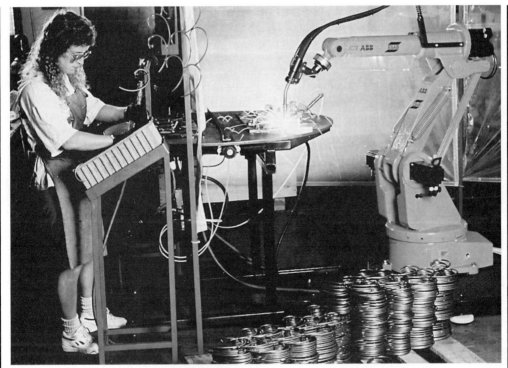

Figure 9–14   Robot Welding of Peacock Tails. (Courtesy of ABB Flexible
Automation)

turing a quality product at an affordable price utilizing the most effective production means and equipment available. Since we purchased ABB's IRB 1500 in 1993, it has proven its ca-pabilities and continues to present an impressive demonstration of technology."

(Courtesy of ABB Flexible Automation)

- *Paint spraying—airless:* The coating material is pushed through a small nozzle opening by a pump in the spray head to atomize the material. The residual momentum from the atomizing process causes the material to coat the part. Servo arms are required when robots are used, and continuous-path control of the spray gun is necessary.

- *Paint spraying—compressed air:* The atomizing action in this system is caused by compressed air passing over a venturi. The residual momentum of the paint and air mixture carries the material to the target. Again, servo-type controllers with continuous-path control are necessary.

- *Paint spraying—electrostatic:* Paint is given an electrostatic charge as it passes through the spray head, and the part is connected to a voltage source with the opposite polarity. The paint is pulled to the part by the difference in the electrical charge between the paint droplets and the part. Continuous-path servo machines are used for robot applications in this area.

Spray painting and coating are well established robot applications. Concerns for the health of workers and gains from productivity force manufacturers to use robots for all large painting applications. Automotive and truck manufacturers rarely use a human in a production paint area. Figure 9–15 shows a coating robot painting window shutters positioned in front of the robot by an overhead material-handling system. A primary advantage of the robot in painting is the repeatability of the programmed motion. If a master painter is used to record the motion of the paint program, then every robot becomes a master painter. Heavy use of painting robots occurs in the automotive and aerospace industries and moderate use occurs in heavy-equipment industries.

Figure 9–15   Coating Robot
(Courtesy of DeVilbiss Company)

CASE: CIM AUTOMATION AT WEST-ELECTRIC

Early in the blade manufacturing process, the W-E team started the development of a discounted cash flow justification spreadsheet (Figure 9–16) that would indicate payback and permit what-if analysis of automation options. A description of the data required for each cell was placed in cell notes in the spreadsheet so that the justification instrument could be distributed to other automation teams in W-E without the need for paper documentation. ●

9-5 SUMMARY

The goal of manufacturing is to make a profit, so justification of capital investments is necessary. In the past, justification of capital expenditures focused primarily on tangible savings and revenue changes. To meet the challenges provided by increased local and global competition, the justification process must include intangible gains that result from the automation process. Tracking these intangible gains is difficult, but transaction-based cost accounting techniques help to uncover cost savings and revenue gains from nonmeasurable sources. Three methods are used most often: payback, return on investment, and cash flow. If the discount rate is added to the calculations, the justification considers the future value of money and the analysis provides the net present value for the justification. Justification of robot applications is enhanced by the long life-time of the machine.

The applications usually associated with robots include material handling, machine tending, assembly, process, welding, and painting. The type of robot used is dictated by the application environment, size and shape of the part, and production rate that must be supported. Robots are used in all major industries; however, the greatest concentrations occur in the automotive, foundry, light manufacturing, electrical/electronic, heavy-equipment manufacturing, and aerospace groups.

QUESTIONS

1. Describe the type of industry environment that permitted justifications to be based only on tangible savings and revenue gains.
2. What forced a broader view of justification of capital investments and the introduction of intangible benefits?
3. Describe the nonmeasurable factors most frequently included in current justifications.

Line Description	0	1	2	3	4	5
				Years		

Capital Equipment

	Line Description	0	1	2	3	4	5
1	Equipment Cost						
	Robot Systems	(110,000)	0	0	0	0	0
	Material Handling	(12,000)	0	0	0	0	0
	Sensors & Safety System	(11,500)	0	0	0	0	0
	System Controller	(8,000)	0	0	0	0	0
	Cell Controls	(11,500)	0	0	0	0	0
	Software	(7,500)	0	0	0	0	0
	Inspection Systems	(15,500)	0	0	0	0	0
	Modification of Ovens	(7,500)	0	0	0	0	0
	Modification of Presses	(8,500)	0	0	0	0	0
2	Freight and Installation	(11,000)	0	0	0	0	0
3	Sale of old Equip.	0	0	0	0	0	0
4	Tax on Old Equip	0	0	0	0	0	0
5	Total Net Investment	(203,000)	0	0	0	0	0

Non Capitalized Cost

	Line Description	0	1	2	3	4	5
	Moving Equipment	(6,500)	0	0	0	0	0
	Installation	(8,000)	0	0	0	0	0
	Training	(15,000)	0	0	0	0	0
	Programming	(18,000)	0	0	0	0	0
6	Total Non-capitalized Cost	(47,500)	0	0	0	0	0

Inventory Change

	Line Description	0	1	2	3	4	5
7	Inventory Change	0	0	0	0	0	0

Operating Costs

	Line Description	0	1	2	3	4	5
8	Direct Labor	79,488	158,760	158,760	158,760	158,760	158,760
9	Indirect Labor	(7,000)	(15,000)	(15,000)	(15,000)	(15,000)	(15,000)
10	Maintenance Costs	0	0	0	0	0	0
11	Tooling Costs	0	0	0	0	0	0
12	Materials & Supplies	500	1,000	1,000	1,000	1,000	1,000
13	Inspection	2,300	6,500	6,500	6,500	6,500	6,500
14	Assembly Cost	0	0	0	0	0	0
15	Scrap & Rework	4,000	9,500	9,500	15,000	15,000	15,000
16	Downtime	0	20,000	20,000	20,000	20,000	20,000
17	Utilities	(200)	(400)	(400)	(400)	(400)	(400)
18	Taxes & Insurance	0	0	0	0	0	0
19	Subcontracting	0	0	0	0	0	0
20	Safety	0	0	0	0	0	0
21	Programming	(8,000)	(2,000)	(2,000)	(2,000)	(2,000)	(2,000)
22	Process Improvements	(4,500)	2,000	2,000	2,000	2,000	2,000
23	Other	0	0	0	0	0	0
24	Total Operating Cost	66,588	180,360	180,360	185,860	185,860	185,860

Other Impact on Revenue

	Line Description	0	1	2	3	4	5
25	Change in Volume	0	5,000	10,000	25,000	25,000	25,000
26	Reduced Lead Times	0	0	0	0	0	0
27	Increased Quality	0	0	0	0	0	0
28	New Product Introduction	0	0	0	0	0	0
29	Manufacturing Flexibility	0	6,500	9,500	15,000	15,000	15,000
30	Other Revenue	0	0	0	0	0	0
31	Total Revenue	0	11,500	19,500	40,000	40,000	40,000

Discounted Cash Flow Analysis

	Line Description	0	1	2	3	4	5
32	Total Operating Cost	66,588	180,360	180,360	185,860	185,860	185,860
33	Total Revenue	0	11,500	19,500	40,000	40,000	40,000
34	Non Capitalized Cost	(47,500)	0	0	0	0	0
35	Total Pretax Cash	19,088	191,860	199,860	225,860	225,860	225,860
36	Tax Rate	0.36					
37	After-tax Cash Flow	12,216	122,790	127,910	144,550	144,550	144,550
38	Depreciation Rate	0.000	0.071	0.143	0.143	0.143	0.143
39	Depreciation	0	(14,413)	(29,029)	(29,029)	(29,029)	(29,029)
40	Depreciation Cash	0	5,189	10,450	10,450	10,450	10,450
41	Inventory Cash	0	0	0	0	0	0
42	Total Net Investment	(203,000)	0	0	0	0	0
43	Total After-tax Cash	(190,784)	127,979	138,361	155,001	155,001	155,001
44	Discount Rate	0.15					
45	Discount Factor	1.00	0.87	0.76	0.66	0.57	0.50
46	Discounted Cash Flow	(190,784)	111,286	104,621	101,916	88,622	77,063
47	Cum. Cash Flow	(190,784)	(79,498)	25,123	127,039	215,661	292,724

Figure 9–16 Upset Forging Justification.

4. Describe the three payback methods.

5. Explain the term *future value of money.*

6. How does discounting affect the three payback methods?

7. What intangibles do robots offer in the justification process that are not found in other types of automation hardware?

PROBLEMS

1. A manufacturer was not satisfied with the irregular cycle time and production levels in the manual production of plastic garbage cans with an injection molding machine. The efficiency of the machine operators varied during the 8-hour shift. The operator ran a part every 43 seconds for the first 2 hours of a shift and every 48 seconds for the middle 4 hours. During the last 2 hours of the shift, the cycle time averaged 52 seconds. This efficiency level was essentially the same during all three shifts. The manufacturer used a robot and achieved a consistent 41-second cycle time for all three shifts. Determine the following:

   a. How many garbage cans are produced by the robot cell over three shifts?

   b. What was the percentage increase in garbage can production with the robot work cell?

2. A manual welding operation with two multi-axes positioning stations assembles the base for a computer mainframe in a two-shift operation. The job requires 44 seams 2 inches long from a variety of angles. In the manual mode, the welding process takes 45 minutes, with 10 additional minutes for setup and inspection. The application was moved to a single robot MIG welding work cell that included multi-axes fixtures for two bases and seam-tracking capability. After the cell was automated with a robot, a base assembly was welded in 11.6 minutes. One operator now supports the cell by loading a raw material package and inspecting finished products. Determine the following:

   a. How many shifts will be required to produce the same number of bases with the automated cell?

   b. How many workers can be reassigned to other production operations in the plant?

   c. The cost of the robot was $65,000 and the 2 positioners were $12,000 each. Cell design costs were 28 percent, programming costs were 12 percent, and miscellaneous costs were an additional 14 percent (costs are percent of robot and positioner cost). The hourly rate (labor and benefits) for workers displaced was $15.50 per hour, the tax rate is 33 percent, the depreciation is 14.3 percent of the cost of the robot and positioners, and the discount rate is 10 percent. Calculate the payback time, ROI, and discounted cash flow using the W-E spreadsheet.

3. At a major industrial enclosure manufacturer, five manual paint spray stations operate two shifts painting the inside and outside of electrical cabinets. The company uses 200 gallons of $30 epoxy paint over the two shifts. It is estimated that 20 percent of the paint is lost by the operator in overspraying

and waste. The company wants to move to automated paint spraying and collected the following data: paint-spraying robot cells cost $150,000 each; waste and overspraying is reduced to 5 percent if robots are used; robots provide a 10 percent increase in productivity; one operator per shift is required for support of the painting cells; the average cost of labor including fringe benefits is $13.50 per hour. Determine the following:

a. How many robot cells are required and what is the cost saving per year in paint?
b. If the tax rate is 37 percent and the depreciation is $21,000 per robot cell, what is the payback time and what is the ROI?
c. Use the W-E spreadsheet analysis to determine payback time using the discounted cash flow analysis. The discount rate is 12 percent.

CASE PROJECTS

1. Assume that you are on an automation team at another W-E plant and are asked to prepare a report describing how the discounted cash flow spreadsheet is used to determine payback. Use the notes attached to all data cells in the spreadsheet and prepare a report that includes the format required for all data entered into the spreadsheet and a description of how the calculations determine the payback period.

2. Your team at the other W-E site decides to include the impact of inflation in the discount cash flow analysis spreadsheet, and you are asked to determine how the equation for present worth (including an inflation parameter) could be added to the spreadsheet calculation. Prepare a report on your investigation.

3. Use the W-E discounted cash flow spreadsheet to determine the payback time for the extrusion work cell developed in question 2 of the case/projects section in Chapter 4. Use a discount rate of 10 percent, a tax rate of 35 percent, and no intangible benefits. Make good financial and engineering assumptions and estimates for input data not directly available from the case study.

4. Identify the intangible benefits that can be used for the extrusion cell justifications, describe how they will be used as revenue entries in the spreadsheet, and compare the payback period when intangible benefits are used with the analysis in question 3.

5. Analyze the payback analysis for the upset forging cell presented in the case study. Determine the following:

a. What intangible benefits did the team include in the payback analysis?
b. Using the notes attached to the cells in the intangible section, describe how the team determined the revenue impact of the intangible benefits.

6. Determine the payback for an upset forging automation design that includes a robot for each manual work cell. Would this have been a better solution?

10

Safety

10-1 INTRODUCTION

In the past, people became acquainted with robots through science fiction writers such as Isaac Asimov. Robots from his writings of the 40s and 50s are especially memorable. From these fictional machines, Asimov developed the Three Laws of Robotics:

1. A robot must not harm a human being, nor through inaction allow one to come to harm.
2. A robot must always obey human beings, unless that is in conflict with the first law.
3. A robot must protect itself from harm, unless that is in conflict with the first and second laws.

These laws are still valid today, and no investigation of robotics would be complete without considering safety issues in their application.

Worker safety is the most important concern; of secondary importance, however, is the fact that damaged equipment causes downtime, which decreases production. It is, therefore, the responsibility of work-cell designers to consider measures to diminish the possibility of accidents to both hardware and humans. In addition, it is the responsibility of the plant safety director to alert workers to the potential hazards of robotic installations.

CAUSE	PERCENT
Incorrect action of the robot during normal operation because of system fault	5.6
Incorrect action or operation of peripheral equipment in the robot work cell during normal operation	5.6
Careless approach of robot by operator, programmer, maintenance personnel, or unauthorized employee	11.2
Human error in teaching or testing	16.6
Incorrect action of the robot during manual operation because of system fault	16.6
Incorrect action of the robot during testing and repairing	16.6
Incorrect action or operation of peripheral equipment in the robot work cell during testing and repairing	16.6
Other causes	11.2

Figure 10–1　Causes of accidents in robot work cells.

A study conducted in Japan on accidents in robot work cells identified the eight causes for robotic accidents listed in Figure 10–1. An analysis of that study indicates that industry must emphasize the following three areas:

1. Effective perimeter warning devices, barriers, and interlocks must be used around work cells with robot-type devices.
2. General training must be provided for all shop floor employees concerning the dangers inherent in robot systems. In addition, comprehensive machine-specific safety training is a necessity for the operators who work in robot work cells.
3. A comprehensive preventive maintenance program must be established for the robot and work-cell hardware by maintenance personnel trained in the hardware and software used in the work cell.

Definitions are provided at the end of the chapter for some of the terms introduced in the text.

10-2 OCCUPATIONAL SAFETY AND HEALTH ADMINISTRATION (OSHA)

The Occupational Safety and Health Act of 1970 represents the foundation for industrial safety standards throughout the United States. OSHA does not have any specific standards relating to robot safety; however, OSHA has issued a directive to field personnel instructing them to use the ANSI/RIA R1506 standard for safety in robot installations. In addition, the OSHA standards in Subpart O, Section

1910.212, detail general requirements for safety precautions around machinery. The major emphasis of this section is on machine guarding, which protects the operator and others from danger at the point of operation (the area where the work is actually occurring).

For nonrobotic fixed machinery, OSHA requires a fixture that prevents operators from exposing any part of their body to the dangers of the operation. The addition of machine guards for human safety within the robot cell, however, may impose severe limitations on the robot's mobility and its access to all parts of the work envelope. One way to view machine guarding is to consider the entire work cell as the point of operation; then the designer can plan methods to protect the workers from such hazards as pinch points, shear points, and collisions within the entire robot work envelope.

A second way to view safety within the work cell is from the perspective of the actual robot production application (welding, material handling, grinding, and so on). Because most current production applications are covered in detail by OSHA regulations, the rules governing a particular application should be reviewed before the design of the robot work cell. Following is a list of pertinent sections:

Subpart N Section 1910.176	Handling Materials—General
Subpart O Section 1910.215	Abrasive Wheel Machinery
Subpart O Section 1910.216	Mills and Calenders
Subpart O Section 1910.217	Mechanical Power Presses
Subpart O Section 1910.218	Forging
Subpart Q Section 1910.252	Welding, Cutting, and Brazing

In addition, special OSHA requirements may be requested at the time of inspection.

10-3 AMERICAN NATIONAL STANDARDS INSTITUTE/ROBOTIC INDUSTRIES ASSOCIATION STANDARD FOR ROBOT SAFETY

This chapter is not intended to duplicate the RIA standard (ANSI/RIA R15.06 Safety Requirements) on robot safety but to provide an overview of the issues and areas addressed in the standard. The major areas covered in the standard are construction, reconstruction, and modification; safeguarding; maintenance; testing and start-up; applications; continuous operation; and training. Safeguarding is addressed in the following section.

10-4 SAFEGUARDING A WORK CELL

The responsibility for safeguarding a robotic cell falls on the user. The standards address safeguarding by considering four points: devices to be used, operator safety, safety of the teacher, and safeguarding maintenance and repair personnel.

Safeguarding Devices

The type, degree, and redundancy of safeguarding must correspond directly to the type and level of hazard present in the robotic cell. The safety plan should include safeguarding devices, barriers, interlock barriers, perimeter guarding, awareness barriers, and awareness signals. The primary function of the devices is to provide a warning and to restrict traffic flow in the area. A strategy for limiting access to the areas is illustrated in Figure 10–2. The work cell is divided into zones 1 to 3, and the zones are defined as follows:

- Zone 1 is the area outside the work cell and has no restrictions on human traffic.
- Zone 2 is the area inside the work cell but out of reach of the robot arm. Only programmers, work-cell operators, and maintenance personnel are allowed in this area.
- Zone 3 is the area inside the robot work envelope, and it cannot be entered as long as the robot is in the automatic or run mode. Penetration is permitted during programming or maintenance but only after safety standards established for the type of robot present have been satisfied.

Zone 1 is typically distinguished by signs and yellow lights. A continuous barrier with electrical interlocks is usually installed to separate zones 1 and 2. Zone 3 is usually distinguished by an outline on the floor indicating the robot's maximum reach. In addition, awareness alarm electronic detectors are frequently used in zones 2 and 3. Some examples of indicators and physical barriers include the following:

- Painted lines on the floor to mark the limits of the work envelope
- Chains and guard posts
- Safety rails
- Wire mesh fencing
- Equipment within the work cell

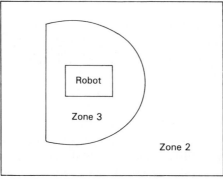

Zone 1

Figure 10–2   Safety zones.

The last type of barrier, the cell equipment itself, when combined with guard rails or fences, provides effective safety protection at reduced cost.

A gate of some type in any physical barrier is required to admit authorized operators and maintenance workers to the robot work area. The standard requires installation of an electrical interlock between the gate mechanism and the robot control. When the gate is opened, either stop the robot and remove actuator drive power to the robot, or stop automatic operation of the robot and any other associated equipment that may cause a hazard.

In addition to physical barriers, protection can also be provided by detectors such as awareness signals, electronic curtains, motion detectors, and pressure-sensitive floor pads. Electronic curtains use visible light, infrared light, or lasers with two or more beams. Laser light can be reflected by mirrors to form a multibeam light curtain. Motion detectors sense the presence of a person in the work area using either microwave or ultrasonic signals reflected by the intruder back to the sensor's receiver. The final type of detector listed uses floor pads with embedded switches that are activated by the weight of the intruder. A typical robotic installation might include a combination of barriers and protection devices.

Safeguarding the Operator

The standards state that the operator safeguards should (1) prevent the operator from being in the robot work envelope when the robot is executing a program, or (2) prevent robot motion when an operator is present in the work space. In addition, operator training should prepare operators for the hazards associated with the programmed robot tasks, so they can recognize them and respond appropriately.

Safeguarding the Programmer

The standards state that the person programming robot operations or teaching translation points should have the necessary training for the programming application, check that hazards do not exist, verify that all safeguards are in place and working, and leave the work envelope before initiating the run mode. When the teach mode is used (1) the robot system and other equipment in the work envelope must be under the control of the programmer; (2) the robot system can operate at high speed under only special operational conditions; and (3) no robot motion can be initiated by remote interlocks or external signals. Finally, only the programmer is allowed in the robot work space when the robot is in the teach mode.

Safeguarding Maintenance and Repair Personnel

The standards address the maintenance of the robot in both the power-up and power-down modes. Rules are provided for the training of maintenance personnel; in addition, rules to cover entry of the work envelope and maintenance of a robot under power are listed.

Because safety consciousness begins with the individual worker, the following guidelines should be emphasized to each worker:

- Rule 1: Respect the robot. Do not take the robot for granted or make an assumption about the next movement the arm will make.
- Rule 2: Know where the closest emergency stop button is at all times.
- Rule 3: Avoid pinch points at all times.
- Rule 4: Know the robot. Pay attention to unusual noises and vibrations from the machinery.

10-5 DEFINITIONS

- *Automatic operation:* The unattended operation of a robot performing programmed tasks
- *Awareness barrier:* A physical device or visual indicator that warns a person of a hazard
- *Awareness signal:* An audible sound or visible light device to warn of a hazard
- *Emergency stop:* A mechanical switch that removes drive power from the robot actuators and causes all moving parts to stop by overriding all other robot controls
- *Interlock:* An electrical circuit or mechanical mechanism that brings about, or prevents, the operation of another device with the operation of the interlock
- *Limiting device:* A device attached to the robot or work cell that restricts the robot work envelope by stopping all robot motion independent of the control or application program

CASE: CIM AUTOMATION AT WEST-ELECTRIC

Early in the W-E upset forging automation project, Bill met with the enterprise team to work on revisions of the corporate safety policy and plan. He was asked to review the safety requirements standard (ANSI/RIA R15.06) and propose a safety plan for the upset forging work cell. After meeting with the design team and several robot vendors, he sent a report on work-cell robot safety to the corporate committee.

SAFETY PROPOSAL—AUTOMATED UPSET FORGING WORK CELL

Operational Overview

The three manual upset forging production cells (Figure 4–32) are modified as illustrated in Figure 4–33 to support automated production. Under the planned operation, one forge is always available for die change and setup while the other two are in production. After a change to a new forge, the previous forge die is cooled for 1 hour, then the die is changed, tested, and preheated. This die change cycle occurs about every 4 hours. The parts hopper is resupplied every 2 1/2 hours, and the inspected parts are collected every 8 hours. Checking and reprogramming of transition points in the work cell occurs about twice a month. As a result of this maintenance, a safety system is necessary that protects the operators and programmers who must work inside the operating cell.

Work-Cell Safety

The automated work cell with safety hardware in place is shown in Figure 10–3. The following safety features are included:

- A flashing amber light above each robot indicates that the robot is in the run mode.
- Two security zones provide for safety. Zone 1 is the area inside the work cell but outside the robot work area, and zone 2 is inside the work area assigned

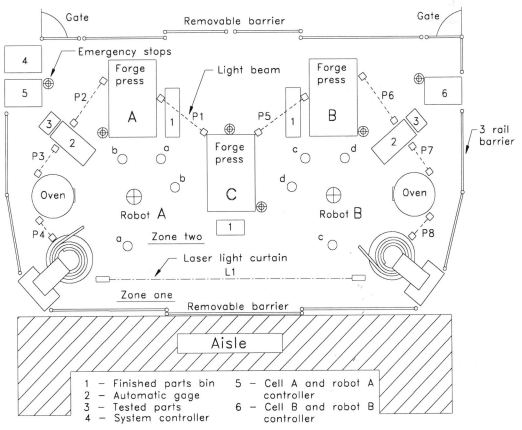

Figure 10–3 Work-cell safety system.

to the robot. Zone 1 has a 40-inch high three-bar rail fence as an access barrier, and zone 2 uses a combination of a laser light curtain and photoelectric sensors to detect unauthorized traffic into the robot work area.

- The zone 1 barrier has breaks on each side of the parts hoppers so that restocking of the hoppers does not require access to the cell. The controllers are also located in breaks in the barrier so that interlock switches used to disable parts of the safety warning equipment are accessible from outside the cell. The zone 2 barrier is composed of eight photoelectric sensor sections and a laser light curtain. Sections can be disabled to allow access to the cell while the robot is operating for maintenance and die change. Lockout switches in the cell controller allow individual sections to be disabled during die change and cell troubleshooting.

- Two gates with position sensors provide access to the rear of the cell for maintenance personnel. The gate sensors can be locked out from the cell controllers to disable the gate-close sensor.

- The zone 1 barrier has removable sections at the front and rear of the cell to allow access for the die change equipment. Each section has a lockout switch on the cell controller that disables the warning system in preparation for removal of the barrier.

- Detection of unauthorized entrance into the robot work area activates an audible alarm and strobe warning light. In addition, the robot is stopped and put in a wait state, and automatic operation of the cell is halted.

- Depressing any of the emergency stop switches around the cell removes power from the robot arm and all cell hardware. Use of the emergency stop switches at the robot and cell controller cabinets removes primary power from the cell. Activation of any emergency stop switch triggers an audible alarm and strobe warning light.

- Work-cell operators do not have to enter the robot work area during the resupply of raw material or the removal of finished products. Only trained maintenance personnel and work-cell programmers have keys to the lockout switches that disable the safety warning and detection system.

- All work-cell operators and maintenance personnel receive training on robot and automated work-cell operation appropriate for their work-cell responsibility.

- Changes in the status or the safety system, such as locking out a safety gate or photoelectric detector for a die change, are collected by the work-cell controller and saved in a log file on the cell controller. The log is uploaded periodically to the area controller by the production control department.

- Maintenance staff changing dies or performing forge setup are protected from injury by the robot with software axes limits and by removable barriers (Figure 10–3). The barriers and soft limits prevent the robot arm from leaving the current work area and entering the area occupied by maintenance personnel changing a forge die. The soft axes limit prevent the robot from moving an axis beyond the limit set in software; the barriers are metal posts inserted into holes in the floor (marked by letters a, b, c, d in Figure 10–3) that physically prevent the robot from entering the maintenance area. ●

10-7 SUMMARY

Safety is an important factor in robotic cell design—from the beginning of the design to the implementation of the process. Specific areas of safety awareness in a robotic installation are addressed in the ANSI/RIA R15.06 Safety Requirements standard. This standard is the OSHA and industry primary reference on robot safety issues governing the design and operation of robotic cells. The standard addresses the following areas: construction, reconstruction, and modification; safeguarding; maintenance; testing and start-up; applications; continuous operation; and training.

Safeguarding considers four issues: devices to be used, operator safety, safety of the teacher, and safeguarding maintenance and repair personnel. The recommended devices include barriers, interlock barriers, perimeter guarding, awareness barriers, and awareness signals. The primary functions of these devices are to provide a warning and to restrict traffic flow in the area. In many applications the robotic work cell is divided into two work zones: inside the cell but outside the work envelope and inside the cell and inside the work envelope. The barriers or markers used to identify these areas include painted lines on the floor to mark the limits of the work envelope, chains and guard posts, safety rails, wire mesh fencing, and equipment within the work cell. When detection devices are triggered, either the robot is stopped and actuator drive power is removed, or all automatic operation of the robot and work cell is halted. Awareness signals are often provided to indicate that a security area has been entered. Electronic curtains, motion detectors, and pressure-sensitive floor pads detect unauthorized penetration.

Operators cannot be in the work envelope when a robot is executing a program, and all personnel responsible for operation of the work cell must have training in robot operation. To protect the robot programmer, the system must transfer control of all work-cell equipment to the programmer during robot programming, move the robot at reduced speed, and not be triggered by any interlocks or external signals. Also, only the programmer is allowed in the work envelope during the point teaching process.

To safeguard maintenance personnel the following rules must be followed: respect the robot, don't take the robot for granted or make an assumption about the next movement the arm will make, know where the closest emergency stop button is at all times, avoid pinch points at all times, and pay attention to unusual noises and vibrations from the machinery.

Asimov's three rules are quite simple but clearly state the requirements for robot safety, namely, protect humans, other equipment, and the robots themselves from harm.

QUESTIONS

1. What are Asimov's three laws of robotics?
2. What is the greatest concern in the work cell from a safety standpoint?
3. What are OSHA recommendations on robot safety?
4. What areas of robot work-cell safety are covered by the ANSI/RIA standard on robot safety?
5. When do most robot-related accidents occur?
6. What rules should be followed to protect the operator and maintenance personnel?

CASE PROJECTS

1. Assume the role of a member of the W-E enterprise safety team responsible for the review and approval of Bill's proposal for the automated robot cell. Write a response to Bill that outlines the strengths and weaknesses in the proposal and list any changes you think are necessary. If a copy of the ANSI/RIA R15.06 standard is available to you, compare Bill's plan with the minimum requirements in the standard and list any deviations or omissions in Bill's plan.
2. Prepare a safety plan for the extrusion work cell developed in earlier chapters.
3. Modify the W-E cell architecture for the upset forging cell in Figure 7–16 to reflect the additional safety systems added to the work cell in Figure 10–3.
4. Modify the W-E cell interface wiring diagram in Figures 7–17 and 7–18 to reflect the additional safety systems added to the work cell in Figure 10–3.
5. As a result of development problems on the automatic gaging system, the vendor building the device requests that the robot place the part into the gage in a horizontal position with the part gripped from the small end of the part. If the robot removes the part from the forge by gripping the smaller diameter of the part just below the upset end, how could the part be reoriented to the horizontal position with the robot holding the part at the end of the smaller diameter?

Human Interface: Operator Training, Acceptance, and Problems

11-1 INTRODUCTION

Automated processes are most frequently implemented because of an anticipated increase in productivity or product quality. The impact of automation is felt by almost every segment of society, from the suppliers of automated hardware to the workers who will be operating the system, to the employees displaced by automation, to the consumers of the products of automation. The plan for implementing an automated system usually includes sufficient time and resources to solve the engineering problems but often overlooks the need for an equal emphasis on the human interface that is present in all such systems.

The following three distinct training activities should be implemented by every industrial plant that considers introducing robots to the production process:

1. General employee awareness training with an overview of the new types of automation, especially robots
2. Operation and programming training programs for all personnel involved in the production process
3. Maintenance training programs for all personnel who will be responsible for maintaining the robot system

The need, type, and target groups for which training must be planned are included in the ANSI/RIA R15.06 Safety Requirements standard. The standard states that the user must provide the following: (1) training for employees who program, teach, operate, maintain, or repair robots or robot systems; (2) training in standard safety procedures and the robot vendor safety recommendations; (3) integration of training in safety precautions and procedures for the specific robots, installation, and application across all project activity.

Plans should be made to have the appropriate phases of the training occur before, during, and after the implementation process.

11-2 GENERAL TRAINING

"Robots are just another form of automation" is a statement that has often been made by those responsible for developing automated manufacturing systems. The statement can be supported from the hardware standpoint without difficulty, but because of the way robots are perceived by both management and labor, training techniques for robot projects must be especially well thought out. Although the difference between robotic automation and automation in general is more psychological than material, it is a potent force in the minds of employees and management. For workers, robots create a higher level of anxiety over possible job loss than do other automated processes, and robots also encounter greater resistance at the management level. Management resistance starts at the level of the first–line supervisor and extends up through varying levels of middle management. Therefore, a well-structured general training program is necessary to deal with personnel concerns about robot automation.

General Training Program

The general training program to prepare all levels of labor and management for the implementation of robots and integrated automation should include the following procedures:

- An overview of current automation practices used by industry in general and competitive industry in particular
- A thorough explanation of the need for broad-based manufacturing automation based on production and marketing data
- A complete description of robot automation including what robots are and what they can and cannot do
- A comprehensive statement of company policy regarding jobs eliminated or changed as a result of automation
- A detailed plan describing how all employees affected by the automation will be retrained
- A clear statement of support from the highest level of management for including automation in any productivity improvement process

Today robots are rarely used alone; they are a part of a work-cell automation system. As a result, training must go beyond just that required for the robot. The training must have a broad focus and address all the computer-controlled hardware that is integrated into the manufacturing system.

The audience receiving this general training is extremely varied, both in the depth of information needed and the emphasis on information delivered. The management level, for example, would profit from a complete overview of current and future robot and system automation for the industry and from training in the specific skills required to implement the integrated robotic cells. The production worker, conversely, would be most interested in the company's plans for workers displaced as a result of automation.

Timing of the training is also important. Starting well in advance of the first robot implementation and automation project is important, but even those companies that currently use robots in automated cells would profit from a general training program before the next installation. Continued information exchange during and after installation is important to keep everyone involved in the current project or future projects up to date.

Training techniques for general audiences are as varied as the groups of listeners they address. General information on robots and CIM may be a topic for an employee newsletter or company paper. Bulletin boards, quality circle groups, shop and department meetings, and seminars can also be effective training media.

The rationale for general training programs of this nature is simply that successful implementation requires a team effort from every worker, and informed workers are better team players.

11-3 OPERATOR TRAINING

The term *robot operator* remains undefined. In one case the human operator may simply activate the work cell at the start of the shift, in other situations the operator works side by side with the robot, and in others a team of workers may be responsible for a broad range of cell activities. The list of responsibilities for a robot operator or cell team may include programming, program loading, work-cell support, and routine system checkout. Currently, support engineering groups do most of the programming and system checkout, and operators perform that portion of their former job that the robot automated work cell cannot do. Operator training related to production in the work cell generally occurs in the plant facilities.

Most robot vendors recommend that training at their facilities begin 2 to 4 weeks before the machine arrives.

11-4 MAINTENANCE TRAINING

To understand the maintenance requirements for robots and automated work cells, it is necessary to review the type of hardware that will be present. Robots and integrated production hardware currently available are powered by hydraulics, pneumatics, or electric drives. The electric drives are both dc and ac servo in equal number; ac drives are becoming the most frequently used on new equipment. The mechanical mechanism itself has the standard gear, pulley, belt and cam linkages, and drives found on automated machines used in manufacturing. Every robot system has a controller, which is either a special-purpose computer or a programmable controller. The special-purpose computers use either microprocessor technology or minicomputer technology as the central processing unit. The robot work cell includes all the manufacturing hardware required for production, one or more robots, sensors, and material-handling equipment. The cells have some type of cell-control computer that is connected to an area controller by a local area network. In many installations the production cells are linked to a corporate host computer and share a single database for all production information.

Maintenance training on the robot, work-cell manufacturing hardware, and the CIM system has generally become a routine procedure. Robot vendors, like most manufacturers of production hardware, have 1- and 2-week training programs designed to train maintenance personnel. The training includes preventive maintenance and troubleshooting of hardware failures. Using the built-in diagnostics in the system, an operator can isolate most of the problems to a printed circuit board so that troubleshooting at the board level is all that is required. If a failure cannot be located, the vendors have field service personnel who will come to a facility and do in-plant repair.

Maintenance training at the system level rests squarely on the shoulders of the end user. The work cell typically has equipment from five or more vendors, and each piece of hardware may contain an electronic control system or computer. The interfacing of the hardware to achieve coordinated operation is usually the responsibility of corporate engineering or a systems engineering company. Because of the unique nature of each package, diagnostic tests are usually not available for use by maintenance personnel, and system failures frequently require troubleshooting to the device level. Development of an effective maintenance training program for the system level in a manufacturing work cell presents the stiffest challenge in the training area.

11-5 TEAM-BASED MANUFACTURING

The dramatic changes in technology and the global market have forced the manufacturing industry to use more work teams. The worldwide market demands a rapid response to customer needs, frequent product introductions, increased productivity, lower production costs, and higher quality. Achieving these *order-winning criteria* with the work force management policies of the 1950s and 1960s is not possible. Many companies have introduced *self-directed work teams* to give them the competitive edge. The definition of a self-directed work team in manufacturing varies. The primary issue is the authority to act and the power to carry the actions to completion. In the purest definition of a self-directed work team, the team is given the broad authority to solve process and personnel problems in their work area. Most companies have not reached that level of employee enpowerment; therefore, the self-directed work team is an expansion of the concept of the *manufacturing process improvement team*.

The following sections describe the concept of the self-directed work team.

Description of a Self-directed Work Team

Several common, generic characteristics describe a work team:

- Frequently a self-directed work team is formed from a functional work group currently in a process area. The term *functional* means that all members of the team come from a single process area.

- The work team meets as a functional group to manage the process area and resolve problems that reduce the effectiveness and profitability of the process. In many organizations the process improvement performed by work teams is an extension of the continuous improvement process initiated for the implementation of CIM automation.

- Team members, trained to work in a self-managed mode, take responsibility of the work area in which they function. Groups are managed by natural leadership from within the group.

- The team follows a five-step process to improve the production process in its area: (1) identification of a process problem, (2) development of a process improvement plan, (3) implementation of the process improvement, (4) tracking of plan effectiveness through data collection, and (5) evaluation of process data to determine performance.

In many companies the team members receive financial rewards from team-implemented improvements in the process.

Making Work Teams Work

The first issue is the *readiness* of the workforce. No amount of training can change a workforce that is not ready to accept the responsibility placed on a self-directed team. Readiness implies that the workers understand the operation of the process and want ownership of it. When asked "Could you do your job better if something was changed?", workers ready for the team process know specifically what needs to be changed to improve the performance of the manufacturing process. On the other hand, workers who are not interested in being a part of the improvement process make poor team members.

Successful teams generally have members who know how the production system functions, have a high level of motivation for improvement in the process, and have a sense of ownership of their jobs. If these conditions are present, the group is ready to be a self-directed work team.

The transition to a self-directed work team is smooth when four rules are followed: (1) management must understand the issues and transfer the authority to act to the functional team; (2) a plan, based on local culture and existing readiness, must be developed for the specific plant location; (3) the team must receive training in self-management skills; and (4) the transition of power to the team must be incremental and spread over a number of years.

Companies using the team concept have an advantage over manufacturers who choose to maintain the traditional supervisor-worker relationship. In organizations where the workers stop thinking when they arrive for work, only a handful of employees are working on the solution to enterprise problems. In contrast, every employee is working on the solution to enterprise problems in a company that successfully adopts the concept of work teams. Over time, the company using work teams will surpass the mindless competition.

11-6 RESISTANCE

Resistance to automation is an age-old problem. An industrial consultant who worked with the General Electric Company on reducing resistance to robots suggests the following three guidelines to help overcome management and worker resistance:

1. Organizations may not install robots to the economic, social, and physical detriment of workers or management.
2. Organizations may not install robots through devious or closed strategies that reflect distrust or disregard for the workforce, because surely they will fulfill their own prophecy.
3. Organizations may only install robots on those tasks that, although currently performed by a person, are tasks that the person performs as a robot would.

Many manufacturers may find it difficult to adhere to the ethics expressed in these guidelines, but maintaining some degree of openness, as in guideline 2, will certainly aid in the implementation of manufacturing automation.

Resistance to robot and manufacturing automation at the management level stems from several causes:

- Concern by middle managers that an unsuccessful automation project will endanger their future promotions
- Fear by first-line supervisors that robot and manufacturing automation will cause additional employee problems and that they will not have management's support in solving them
- Concern by first-line supervisors that production output will drop during the transition to automation and that they will be held accountable
- Conflicts concerning responsibility for the project, along with jealousy between manufacturing units and disagreements about distribution of resources

Resistance to robot automation at the labor and management levels arises basically from a fear of the unknown. This could be fear of losing a job, fear of being transferred, or fear of the demands of retraining. Much of the resistance resulting from these fears and those of managers can, however, be neutralized by an effective training program at the three levels defined earlier.

11-7 ORGANIZED LABOR

Because organized labor believes that new union members do not want to do many of the tasks that are dirty, dangerous, or dull, it now officially welcomes robots and manufacturing automation systems. In addition, labor realizes that increased productivity is necessary. Organized labor, however, stands firm on two demands: (1) displacement of workers must be gradual, and (2) labor should get a percentage of the benefits that result from robot automation.

Labor acknowledges the need to remain competitive with foreign production through implementation of enterprise-wide automation, but it is reluctant to give management a free hand in the implementation. Unions still fear large-scale displacement of blue-collar workers as a result of automation. Unions are based on the principle of solidarity among workers, and as the ranks decline so does the operating revenue of unions. Job security and training continue to be the major issues in most contract negotiations. Through the end of this century, union membership drives will concentrate on the growing number of white-collar workers, especially those involved with manufacturing automation.

CASE: CIM AUTOMATION AT WEST-ELECTRIC

The performance of the upset forging cell was measured regularly after the automation was fully functional. At the 6-month mark, the following performance measures were recorded:

- Cycle times of 13.5 seconds per part are maintained in the operation of all forges.
- Cycle times of 9 seconds were demonstrated with forges A and B.
- Setup and die change time was reduced 40 percent.
- Lot sizes of 1000 parts can be produced economically.
- Die life has increased 25 percent.
- The number of upset forged parts that exceed the tolerance is reduced by 60 percent.

11-8 SUMMARY

Training is a critical component in every CIM implementation. The training process should include general employee awareness training, operation and programming training for the technical staff, and maintenance training on robots and automation hardware for the maintenance staff. This type of training is required by the ANSI/RIA safety standard.

The general training should include an overview of current automation practices, a rationale for the introduction of automation, an explanation of what robots are and how they work, company policy regarding introduction of automation and displacement of workers, retraining planned for displaced workers, and a statement demonstrating support from top management. The timing, level, and emphasis for the training must be matched with the target audience.

Operator training should start 2 to 4 weeks before the robots are in place and include a level of detail that is consistent with the responsibility of the operators in the automated cell. The content of the maintenance training should be consistent with the type of robot system and automation hardware planned for the cell. Training on the individual pieces of automation in the cell, such as robots, programmable logic controllers, and computer numerical control machines, is usually provided by the vendor selling the hardware. The training for the system operation of the cell is the responsibility of the company performing the integration. In many cases the manufacturer must conduct the system training to achieve the best results.

Work teams are being developed daily in the manufacturing industry. The level of authority to make changes in the process and to manage the work area indicates the degree of self-direction in the team. Implementation of work teams requires a workforce that knows the operation of the manufacturing process and that wants to take ownership of the system. Companies implementing self-directed work teams swell their ranks with additional problem solvers and increase their competitive edge in the marketplace.

Resistance to automation is not new. To overcome worker resistance, companies should use automation for the benefit of employees, communicate about automation issues openly and frankly, and install robots to free workers of dull, dangerous, and dirty jobs. Overcoming management resistance to automation requires an understanding of the roots of resistance, for example, concerns that unsuccessful automation projects will have adverse effects on careers, too little management support when automation causes worker unrest, concern that supervisors will be held responsible for transition problems in automation implementations, and turf battles and jealousy over selection of automation projects.

In the past, organized labor has provided only lukewarm support for robot and manufacturing automation. Now the well-documented challenges of the global economy have convinced labor that labor should be part of the solution rather than part of the problem. The issues are speed of automation implementation and the payback to the worker when automation is installed.

QUESTIONS

1. What are the three training activities required in every industrial plant planning to use robots? What do the standards say about the training needs?
2. Describe the elements that should be present in a general training program.
3. What is the primary focus of a general training program for each level within an industrial organization?

4. List the major elements of an operator training program.
5. What complicates maintenance training procedures on automated systems?
6. What is the difference between maintenance at the system level and maintenance at the machine level?
7. What are the key internal and external issues affecting the management of the workforce?
8. Describe the common characteristics of a self-directed work team.
9. What are the three laws for reducing resistance to automation?
10. What causes resistance to robot automation at the management level?
11. Give organized labor's view of automation and robotics.
12. How is the training requirement effected by the type of robot selected, level and number of sensors, and system architecture?

Index

Q

Quality
  definition, 11
  inspection, 197
  as a performance measure, 83
  standard, 10

R

Radio frequency tags, 199
Reasons for rapid development, 4
Rectilinear coordinate system, 36
Reduction-drive, 52
Reference frames, 249
Reliability, robot, 4, 30
Remote center compliance (RCC), 133
Repeatability, robot
  definition, 30
  pneumatic robots, 51
Repetitive manufacturing, 17
Residence time, inventory, 12
Resistance to automation, 306
Resolver feedback, 60
Return on investment, 4, 268
Robota, 2
Robot
  arm, 19
  assembly, 131
  automation survey, 112
  basic system, 19, 20
  classification, 36
  controller, 22, 219, 223
  definition, 13
  geometries, 19, 36
  grippers, 116
  Institute of America, 3
  ISO classification, 74
  jointed-spherical, 20
  power sources, 22
  primary advantages, 14
  production tooling, 21
  specifications, 26
  standards, 32
  system integration, 209
  systems definition, 19
  teach stations, 24
ROI, 4, 268
Roselund, Harold
Rotational coordinates, 29

S

Safety, 291
  accident, causes, 292
  cell personnel, 295
  definition of terms, 296
  devices, 294
  guidelines, 175
  standards, 292
  three laws of robotics, 291
  W-E case, 296
  work-cell, 293
SCARA, 46, 48
Scheinman, 3
Seiko robots, 31
Sensors, 151
  analog, 152
  applications of, 172
  characteristics of, 152
  contact, 153
  discrete, 152, 153
  limit switches, 153
  non-contact, 160
  photoelectric, 167
  process, 173
  proximity, 161
  reasons for use, 151
  tactile, 158
  vision, 103
Sensors for active compliance, 132
Servomotors, 52
Servo-system, 55, 220
Setup time
  definition, 11
  as a performance measure, 83
  standard, 10
Shakey, robot, 3
Simple touch, tactile, 158
Simplification, process, 81, 108
Sony robots, 4
Speed, robot, 31
Special purpose gripper, 128
Special purpose tools, 130
Spherical coordinate system, 36, 44
Spherical geometry, 44
Standards, manufacturing
  Coopers and Lybrand, 10
  world, 13
Standards, robots, 32
Stanford arm, 3
Stepper motor, 52
Stop-to-stop control, 67, 68
Superficial improvement, 82